Gilbert R. Redgrave

Calcareous Cement

their nature and uses, with some observations upon cement testing

Gilbert R. Redgrave

Calcareous Cement
their nature and uses, with some observations upon cement testing

ISBN/EAN: 9783337406288

Printed in Europe, USA, Canada, Australia, Japan

Cover: Foto ©Andreas Hilbeck / pixelio.de

More available books at **www.hansebooks.com**

CALCAREOUS CEMENTS:

THEIR NATURE AND USES.

CALCAREOUS CEMENTS:

THEIR NATURE AND USES,

WITH

SOME OBSERVATIONS UPON CEMENT TESTING.

BY

GILBERT R. REDGRAVE,

ASSOCIATE OF THE INSTITUTION OF CIVIL ENGINEERS; TELFORD GOLD MEDALLIST;
AND OFFICIER DE LA LÉGION D'HONNEUR.

With Diagrams.

LONDON:

CHARLES GRIFFIN AND COMPANY, LIMITED;
EXETER STREET, STRAND.

1895.

PREFACE.

THE marked improvements effected in recent years in the use of limes and cements date back undoubtedly to the close of the last century. They may be ascribed, partly to the great advance made about that time in chemical and scientific knowledge, and the accurate habits of thought and practical research induced by the spread of that knowledge, partly, also, to the rapid growth of engineering skill, which characterised that period.

In both of these directions our own countrymen yielded important services, and, so far as limes and cements are concerned, while the "ingenious" Dr. Brindley Higgins was about 1780 laying the foundations of a proper study of the chemistry of the subject, Smeaton had already, as early as 1756, observed the notable fact that the hydraulicity of limes was due to the presence in them of clay.

The records of the English Patent Office do not, however, disclose any great activity in these branches of invention until thirty or forty years later, and considerable progress appears to have been made in the manufacture of artificial limes abroad, before the subject attracted serious attention in England. The discovery by Parker of the valuable cement made from the septaria of the London clay, patented by him in 1796, was doubtless an invention of the utmost importance, as I shall attempt to show, when dealing with the subject from the historical point of view.

Little injustice will probably be done to those who have turned their attention to these materials, if we state that

from the times of the Romans until Smeaton's discovery—
say for a period of close upon two thousand years—no
substantial advance was made in the employment of limes
and cements. It has, in fact, been reserved for the latter
half of the nineteenth century to demonstrate the superiority
of artificial cements over the best descriptions of natural
cements and admixtures of fat or pure limes with other
substances imparting to them cementitious properties.

Notwithstanding the excellence of the cements now being
placed upon the market, and the amount of attention which
the manufacture and employment of cements have recently
received, we have still much to learn respecting the
chemistry of cement action, and concerning the problems
involved in the ecomomical production of these materials.
On the subject of testing cements we are, be it confessed,
lamentably deficient, as compared with many of our continental neighbours, and in this direction, more especially
as regards the introduction of some uniform and generally
accepted system of testing, there is much still to be accomplished in England.

It has seemed to me, therefore, that a brief account of
the history of the employment of cement in the past, which
has received but little attention from previous writers,
together with a description of the various processes of the
manufacture, might be of use at the present time. I have
also collected some of the opinions of experts on the testing
of cement, and I have appended a full translation of the
German Rules for Testing, which, I think, have not previously been published in this country. These rules would
appear to possess peculiar interest at the present time, in
consequence of the controversy which has arisen among
English manufacturers, even while this work is passing
through the press, concerning the nature of Portland
cement. The question of the legality or otherwise of the

mixture with the cement, during the process of grinding, of less valuable ingredients was fought out in Germany more than twelve years ago, and after careful consideration and many experiments, it was decided both by manufacturers and engineers to adopt such a definition of "Portland cement" as to exclude all such admixtures, except two per cent. of plaster of Paris, employed for a special purpose. With the reasons for this use of plaster I have dealt fully in the ensuing pages, and the question whether Kentish ragstone or any other variety of unburnt material, or slag, should be employed by the manufacturer, would seem properly to come within the provisions of the Adulteration Acts, when we have such an extension of these Acts, as is, I believe, contemplated.

I cannot claim to have dealt with the subject exhaustively, but I have endeavoured to represent the present aspect of the cement question impartially, not only from the point of view of the manufacturer, but also from that of the cement user, in both of which directions I have had the advantage of a practical acquaintance with the facts to which I refer.

The leading works on limes and cements, both in this country and abroad, have been freely consulted and laid under contribution, and I have endeavoured in all cases to acknowledge my obligations to the writers whose opinions I have quoted. I have availed myself extensively of the *Minutes of Proceedings of the Institution of Civil Engineers*, which contain some of the most valuable essays on Portland cement in existence, and I have here to tender my special thanks to my colleague, Mr. Charles Spackman, F.C.S., who has prepared the greater part of the Chapter on Grinding Machinery, and the original and important Chapter on the Chemical Analysis of Cements and Cement Materials. Many other manufacturers have likewise aided me largely with facts and figures, and I wish also to record my indebtedness to Mr. I. C. Johnson for much valuable information.

I trust that this account of what has been achieved in the past, and this brief record of the present state of our knowledge of limes and cements may be useful to those who may come after, and to all engaged in this branch of industry.

I regard the present time as, in many respects, a very critical period in the history of the cement trade. Our own country, the original seat of the manufacture, has been distanced in certain directions in consequence of the superior scientific skill [and the energy of foreign rivals. The supremacy we have so long enjoyed has undoubtedly been to some extent wrested from us by the products of Continental industry and enterprise, and in the absence of some united action and some intelligent leading, our manufacturers are threatened with a competition which they are not adequately armed to encounter.

It would be idle on my part to deny that I dread times of even greater depression than those we have hitherto experienced in this important industry, and I am convinced that nothing but the formation of a powerful confederation of cement producers and cement users, such as exists in many of the most enlightened countries abroad, will extricate us from the difficulties by which we are surrounded. I trust that it may be possible to induce the Institution of Civil Engineers or some influential Corporation to undertake these duties, and to free us from the dangers which I have attempted to indicate.

GILBERT R. REDGRAVE.

THE ELMS, WESTGATE ROAD,
    BECKENHAM, KENT.
        December, 1894.

CONTENTS.

CHAPTER I.

Introduction.

PAGE

CHAPTER II.

Retrospective and Historical Review of the Cement Industry.

CHAPTER III.

The Early Days of Portland Cement.

CHAPTER IV.

The Composition of Portland Cement.

CHAPTER V.

Processes of Manufacture.

CHAPTER VI.

The Wash-Mill and the Backs.

CHAPTER VII.

Flue and Chamber Drying Processes.

CHAPTER VIII.

The Calcination of the Cement Mixture.

CHAPTER IX.

The Grinding of the Cement.

CHAPTER X.

The Composition of Mortar and Concrete.

CHAPTER XI.

Cement Testing.

CHAPTER XVI.

Specifications for Portland Cement.

APPENDIX A.

APPENDIX B.

APPENDIX C.

APPENDIX D.

APPENDIX E.

APPENDIX F.

APPENDIX G.

APPENDIX H.

LIST OF ILLUSTRATIONS.

CALCAREOUS CEMENTS:

THEIR NATURE AND USES.

CHAPTER I.

INTRODUCTION.

Antiquity of the Use of Lime.—The practice of employing limes as binding agents for stones, bricks, and other similar materials used in construction is one of great antiquity, and the Romans, the most mighty builders of the ancient world, were well acquainted with the use of limes, as also with certain of the methods of improving fat or pure limes, in order to impart to them cementitious and hydraulic properties.

Definition **of Cements and Limes.**—"Cements," as distinguished from limes, are materials which are capable of solidifying when in contact with water without perceptible change of volume, or notable evolution of heat; "hydraulic cements and limes" are such as possess the power of "setting" or solidifying under water. All limes have a tendency to expand and to fall abroad, or to crumble into powder when treated with water, and are said to become "slaked"; the purer the lime the more energetic and rapid is this action, while conversely the greater the quantity of clayey matter combined with the lime, the less intense, as a rule, is the chemical affinity for water, and the slower is the act of hydration, and to this extent the greater is the resemblance of such limes to cements. "Limes," therefore, as distinguished from cements, "fall" or crumble when exposed to the action of water.

It may be as well to point out here that the above definitions, though they do not entirely accord with all the previous theories of limes and cements, are those which will be adopted in the present work.

Combination of Lime with Water.—The chemical affinity

1

of lime for **water is one of the** most powerful with which we are acquainted, and "quicklime" (calcium **oxide), or lime** recently calcined, when **exposed to** the air **speedily attracts moisture** from **the** atmosphere, and **combines with such water to form** calcic hydrate, or slaked **lime.** This **hydrate may occupy as** much as three times the space previously filled by **the quicklime,** and therefore the amount of slaked lime produced from a **given bulk** of quicklime **appears in certain** cases **to be** very **considerable.**

The Speed of Slaking.—Some **writers have attempted to** classify **the different** varieties of lime **in accordance with the quantity of** slaked lime produced, or with the speed **with which they were** observed **to** combine with water; but it is now known **that this** slaking **action** depends upon numerous conditions which **have to be** specially studied **for each class** of limes, and that any general **deductions** founded upon the **act** of hydration alone are likely to **be inaccurate and** misleading.

Intermediate Limes.—Certain impure **limes,** resembling in their composition the constitution of **cements, have** been appropriately named "intermediate limes," or such as occupy a position intermediate between the **true limes,** which undergo disruption when exposed to the action of water, and the cements which do not apparently become changed when so treated.

It may be assumed that limes of every different degree of energy, from pure oxide of calcium down to true calcareous cements, exist in nature; thus there is an enormous range of varieties of action to be studied, and any attempt to classify all **limes** under two or three sub-heads must be futile and un**trustworthy.**

In order to rightly understand the action of limes and cements, a certain amount of chemical knowledge is involved, and this branch of the subject may now be briefly discussed and treated of as tersely as possible.

Quicklime **or Caustic Lime.**—Quicklime, caustic lime, or the oxide of calcium, one of the earthy metals, does not exist in nature, nor is metallic calcium itself anywhere found in an un-combined form. We obtain quicklime, the chemical symbol for which is CaO, by calcining or heating to redness a carbonate of lime ($CaCO_3$), and by this means expelling the carbonic acid gas or carbon dioxide (CO_2) with which the lime is combined, and which can be driven off in the gaseous form at a cherry-red heat, about 440° Centigrade.

Preparation of Quicklime.—Lime combined with carbonic **acid** is found in a great variety of rocks in all parts of the world, and in every different degree of purity. From any pure carbonate of lime, such as marble, or any of the various descriptions of

limestone and chalk, the calcium oxide or quicklime can thus
readily be prepared ; but in order to keep the oxide unchanged,
it must, as already stated, be preserved from **contact** with the
atmosphere.

Chemical Composition of Quicklime.—In a pure carbonate
of lime 44 parts by weight of carbon dioxide or carbonic acid are
combined with 56 parts by weight of calcium **oxide.** In the
oxide itself 40 parts by weight of metallic calcium (Ca) are combined with 16 parts of oxygen gas (O). This oxide cannot be
decomposed by heat. The **lime burner** has, therefore, in his
kilns to drive off **the combined carbonic** acid in **the form of a**
gas, and thus to **obtain the lime as used** by **the** builder.

Calcination of Limestone.—Generally speaking, **the limestone or chalk, when placed in** the kiln, contains a certain
percentage of moisture, which has also **to** be expelled, and thus
the lime burner can rarely, when the stone **is** thoroughly well
burned, **and all** the carbon dioxide is expelled, obtain more than
half its weight of quicklime from a given weight of stone dealt
with in the kiln, though in **theory** the yield should be 56 per
cent. of lime.

Carbonic acid gas is driven **off** much more freely in **an**
atmosphere of steam, and for this reason some authorities have
advised **the use** of the steam jet in lime kilns, but we are not
aware that this plan has met with any measure of practical
success.

Action of Aqueous Vapour.—The action of aqueous vapour
upon the lime **undergoing** calcination may be thus explained :—
A vapour or gas **ascends into a** space already occupied by another
gas differing from it in composition, **almost as** quickly as it would
into a vacuum, owing to the law of the diffusion of gases, which
teaches us that the various gases have such a powerful tendency
to permeate each other, that, **in order** to do so, they will frequently overcome considerable resistance. Water, for example,
evaporates as readily into dry air **as** into a vacuum, but this
tendency **is** impeded when the air above it is already charged
with aqueous vapour. In the same way when the red-hot lime
is giving off **carbonic acid** gas, this gas escapes much more freely
into an atmosphere **of aqueous vapour than into one of** carbonic
acid gas, for in this latter case the escaping carbonic acid has to
overcome the pressure of that which has been already evolved.
It is true that if we could substitute a current of air at a very
high temperature for the aqueous vapour the result would be
the same, but this would entail many difficulties.

Gay-Lussac's Experiment.—Gay-Lussac, the eminent chemist, devised an ingenious experiment to demonstrate the effect

of steam upon the calcination of lime. He introduced into a hard glass tube, inserted in a furnace which enabled it to be raised to the required temperature, some pieces of broken marble. One end of the tube was connected with an apparatus for the evolution of steam, and the other end was furnished with a contrivance for collecting the carbonic acid gas. The temperature was then raised to the point at which the marble began to be rapidly decomposed, and at this stage by checking the draught the mass was reduced to a dark red heat, so that all evolution of carbonic acid ceased. Steam was at this moment permitted to pass through the tube, and at once carbonic acid gas again made its appearance in considerable quantities and continued to pass off under these circumstances in a manner entirely dependent upon the current of vapour. This experiment proves that lime-stone is decomposed at a far lower temperature with the help of aqueous vapour than under ordinary circumstances. In an atmosphere of carbonic acid gas, carbonate of lime, as we know from the investigations of Faraday and Hall, remains unchanged even at a full red heat.

Composition of Slaked Lime.—When lime becomes "slaked" it is found that 56 parts by weight of quicklime combine with 18 parts by weight of water (H_2O) to form 74 parts of calcic hydrate (CaH_2O_2). Great heat is evolved in this process, and the action is expedited by the use of boiling water. Certain "poor limes" which will scarcely slake or fall to powder when cold water is employed, will crumble into dust readily if the water is at the boiling point.

Classification of Limes.—Limes are frequently classed as "fat" or rich limes, if they readily become slaked and furnish a large volume of powder, and "poor" if they are impure and become slaked slowly, yielding relatively but little dust. Cases occur, of course, when the lime will not fall or slake at all. Certain poor limes, which are practically natural cements, may, if they are calcined at a proper temperature and are finely ground, be employed as cements. To render these limes avail-able for the use of the builder, it is thus necessary that they should be ground previous to use.

Condition of Water in Slaked Lime.—The water which combines with the lime in the act of hydration is truly solidified, and the hydrate formed is, when the exact proportion of water necessary for this purpose has been employed, an absolutely dry powder. On adding a further quantity of water, the bulk of this powder is much reduced, and it may be tempered into an ex-tremely rich and unctuous paste. If this paste is permitted to dry, it shrinks and forms a porous mass of no great hardness.

Dry Slaked Lime kept from Air remains unchanged.— If the hydrate of lime is preserved from atmospheric action and placed in a hermetically sealed vessel, either as a dry powder or made up with water, it undergoes no alteration whatever, the powder remaining as such for an indefinite period, and the paste showing no sign of change. The hydrate of lime possesses in fact no inherent power of solidification. In proof of this Alberti states that some lime taken from a mortar-pit in an old ditch, abandoned for upwards of 500 years, was "still so moist, well tempered, and ripe that not honey or the marrow of animals could be more so."

Lime slowly recombines with Carbonic Acid.—When exposed to the air, pure caustic lime is converted very slowly and without notable increase of temperature into a rather coarse powder. It is not, under these circumstances, wholly converted into a carbonate of lime, even after the lapse of many years, but, by the simultaneous absorption of moisture and carbonic acid, it is resolved into a double-compound having the formula, according to Fuchs, of $CaCO_3 + CaH_2O_2$, or consisting of equal equivalents of the carbonate and the hydrate of lime. The carbonate thus produced would seem to result from the decomposition of the first-formed hydrate, for when moisture is wholly excluded no combination between the lime and the dry carbonic acid gas takes place. In order to expel the water of hydration, the slaked lime must again be heated to dull redness.

The Action of Carbonic Acid mainly Superficial.—Lime made from pure carbonate of lime, when slaked and used for mortar, likewise gradually recombines with the carbonic acid gas present in the atmosphere and becomes indurated, but this action is mainly in the superficial layers of the mortar, as the gas penetrates very slowly. In fact years must elapse before the recarbonisation of the lime is thoroughly accomplished, and in the case of thick walls the internal layers of mortar never become completely hard. It is necessary to distinguish between the so-called "set" of the mortar, which is merely due to the absorption of the superabundant water, and the actual induration by means of the carbonic acid gas which is a process of years or ages in the case of pure limes.

The Influence of Clayey Matters.—But absolutely pure limestones are only met with in exceptional cases, as nearly all limestone rocks, and the greater part of the chalk formation contain varying percentages of clayey matters (silicates of alumina), iron, alkalies, &c., and it is upon the proportion of these ingredients present that the behaviour of the calcined lime principally depends. It is, in fact, owing to the presence of certain of these clayey matters, first, as we shall see pointed out by Smeaton, that limes pass over by gradual stages into the form of cements; that is to say, that these substances so far influence the slaking action that they may even bring about the ultimate setting of the mixture without change of volume—the characteristic property (as already stated) of cements.

Artificial Admixture of Clayey Matters.—It is not necessary, however, that the limestone should have been the source from which these clayey matters were derived; they may be conveyed to the calcined lime by admixture with it at the time when it is treated with water, or they may be ground up along with **the** lump lime before it is slaked. It is this fact which needs careful consideration when we have to deal with the influence of heat on mixtures of lime and clay, and the nature of the changes effected in the kiln. The silica compounds are of a very complex character, and may be produced, as we shall see, both by heat and in the humid way. All that is necessary for the due action of these clayey matters is that they should themselves have been roasted or calcined, either artificially or by volcanic heat.

Puzzuolana, Trass, &c.—Certain of these substances which are added **to** pure limes to **bring** about this action are called puzzuolanas **or** trass. These **are** clayey or siliceous matters of volcanic origin, but roasted shales, brick dust, and burnt clay or ballast, all of them, more **or** less, possess this influence on the pure limes, and have the power of imparting to them the attributes of cements.

The volcanic ash found **in** the island of Santorin, and known **as** Santorin earth, is typical of many kinds of scoriæ which have been used successfully with fat **or** pure limes to impart to them **hydraulic** properties. **The** proportion of silicate of alumina in **this** substance is relatively high, and **there is** much less iron than in the case of trass and puzzuolana.

Silica and its Compounds.—It will be necessary, in order **to understand the** chemistry of cements, to treat next of silica and its **compounds.** Silica, the oxide of the element silicon, is found very widely distributed in nature, sometimes pure but more often in combination with other substances, as it has great avidity **to form** complex salts, known as compound silicates. **It** plays the part of an acid, and combines with lime, alumina, iron, and the alkalies in a vast number of different proportions. It is found that 28 parts by weight of silicon (Si) and 32 parts by weight of oxygen are present in silicic acid or silica, having the chemical formula of SiO_2. Clay, the silicate of alumina, may be taken as the type of the silica compounds, while quartz, flint, and chalcedony consist of almost pure silica. Porcelain clay, which contains about 47 parts per cent. of silica, 39·2 parts of alumina (Al_2O_3), and 13·7 parts of water, and corresponds to the chemical formula $Al_2O_2 2SiO_2 + 2H_2O$, with a molecular weight of 258·4, may represent the silicates. There are, however, an enormous number of clays in which silica and alumina are present in very varying proportions, and which contain in addition iron, alkaline matters, lime, &c. For certain of these clays it becomes almost impossible to propound any reliable chemical formula, to express their composition; and alumina, while it may combine in certain definite proportions with the silica as a base is also capable of acting as an acid, and of combining with lime and the alkalies to form certain more or less unstable and little known compounds termed aluminates.

Alumina is the oxide of the metal aluminium (Al), which has the atomic weight of 27·2, and two parts of aluminium combine with three parts of oxygen equal 48 to form its only known oxide, termed alumina, amounting in all to 102·4.

It will not now be necessary to study in detail the combinations of silica and alumina with iron and the alkalies—soda and potash—though these compounds play a very important part in cement action; we shall have to revert to this question hereafter when we consider the chemical changes consequent on the hydration and induration of cements of various kinds.

Restricting ourselves for the present, therefore, to the silicates of alumina as typical of the compounds of silicic acid, we may pass on to discuss the modifications brought about in the slaking action of limes consequent upon the presence of these substances. We will take first impure limestones, or such as contain a proportion of the silicate of alumina, together with certain percentages of iron tending to discolour the lime.

Influence of Heat on the Silicates.—When limes of this character are burnt in the ordinary way in the kiln, the carbonic acid gas is first expelled from them as in the case of the pure limestones, and the clayey matters assist in its expulsion, owing partly to the affinity of the silicic acid for the lime and partly to the fact that the free and combined water in the clay is driven off, and the steam produced in this way facilitates, as already stated, the expulsion of the carbonic acid. There is thus a double change to be effected in the kiln, and the expulsion of the water from the hydrated silicate of alumina in the clay may go on side by side with the dispersal of the carbonic acid.

The Calcination of Clayey Limestones. — These clayey limestones are thus burnt more readily than the pure limestones, they also require less fuel and less time. But when the heat is sufficiently prolonged, a further action takes place in the kiln, the silicic acid and the lime react upon one another, and

either produce a new compound, termed silicate of lime, or so
far approximate to the formation of such a silicate that on the
addition of water this compound is formed with great rapidity.
The silica combined with alumina and iron in the clay is free to
pass to the lime as soon as the carbonic acid is expelled, and no
doubt portions of the silica are seized upon by the red-hot lime,
which is in what chemists term the "nascent state," and is,
therefore, in a condition eminently well fitted to form fresh
chemical compounds.

Action of Water on Silicates formed in the Kiln.—In
using for structural purposes limes of this character, therefore, a
fresh set of affinities comes into play. We have, first, the affinity
of the quicklime for water, and, secondly, the affinity of the
hydrate of lime and the silica and alumina to enter into combination. It must be pointed out that this second reaction can
and does greatly change the violence of the affinity of lime for
water, and that it is possible to so "deaden," if we may thus
term it, the action of quicklime that, upon the addition of water,
instead of heating up and falling into powder, it may pass (as we
have seen) into the condition of a hydrated silicate, without
perceptible heat, and with little or no change of physical
condition.

Stable Compounds formed by Hydration.—The hydrated
silicates of lime thus produced are not liable to be altered to any
appreciable extent when exposed to the prolonged action of water
or of the atmosphere, and the compounds of lime with silica and
alumina thus afford a durable cementing material, suitable also
for hydraulic purposes. It must not be forgotten that the lime
which enters into combination with the silicic acid passes from
a slightly soluble into an almost completely insoluble condition,
and it will be observed that the action of induration differs
entirely, in the case of cements, from the slow and tedious hardening of common lime mortar, due to the gradual recombination
of the lime with the carbonic acid gas present in the atmosphere.

Character of Kiln Changes.—It will have been noticed that
we were careful to state that the lime and the silica either entered
directly into fresh chemical combinations, or that they approxi-
mated to the formation of silicates, for the changes brought about
in the kiln, and the subsequent action which takes place when
water is added to the calcined materials are very complex and
difficult to investigate. It is evident from very simple tests
that, even in the case of cements of the Portland type, considerable quantities of lime remain in an uncombined state, or one in
which such lime is free to be acted upon or to undergo further
chemical change. In a well burnt sample of intermediate lime

the silica has assumed that form in which it is soluble in a boiling solution of carbonate of soda, after treatment with hydrochloric acid, for clay which has been calcined in contact with lime or other metallic oxides is readily attacked by this acid, and the silica liberated assumes the gelatinous condition, or that in which it can freely combine with the alkalies. During the calcination, therefore, of hydraulic limes a stage is reached when, after the expulsion of the carbonic acid, the silica and the alumina of the clay on the one hand and the lime on the other are in a condition to mutually react upon one another, and to form compounds capable of being re-arranged in the presence of water. We are not able to state very definitely the nature of these combinations, but this fact is certain that at this period of the calcining process a very slight increase of heat suffices to fuse or partially vitrify the mass, in which case complete combination between the lime and silica has undoubtedly taken place, and the silicates produced in this way by heat manifest but little tendency to undergo any change when treated with water. It is, therefore, most important not to overburn or, as it is termed, to "clinker" these intermediate limes.

Action of **Impure Limestones when** Calcined.—Some clayey limestones are much more liable to fuse than others, and it has been pointed out that those in which the proportion of iron and the alkalies is large are most exposed to this risk. The glass-like or vitreous silicates formed in this way by heat are generally produced only on the surface of the lumps of overburnt lime, and unless such lumps are carefully picked out and cast aside when the lime is being made into mortar, they may be the source of much subsequent injury to the work, first by the retarded slaking of particles of lime encased in the inert silicates, and second by the more gradual hydration of certain of these complex silicates which may be attended by slight increase in bulk, tending to the disruption of the mass.

Limes vary widely in their behaviour with Water.—It will be understood that, between the pure or fat limes, which slake in the ordinary way, and the true cements which set without change of volume, there is an infinite number of varieties of limes which incline in one direction or the other towards these two extremes. There are also numerous hydraulic limes which first fall to pieces during the act of hydration, and which subsequently set and become indurated when the effect of the silicates comes into play. To this category belong many of the limes of the lias formation, and certain of the grey limes of the lower chalk.

Definition of the Energy of Cement Action.—The energy of a cement depends upon the rapidity with which the lime and the silica combine in the presence of water to form stable compounds, or with which the ready-formed silicates become hydrated when water is added. We have thus the quick-setting cements of the Roman cement type, which become indurated mainly by simple hydration in a few minutes, and the dense cements resembling Portland, which depend for their induration on a re-arrangement of the silicates, and which may take as many hours to set as the former substance does minutes.

It should be here noted that when we speak of the setting of cements we imply the act of induration and not the mere absorption of the water, which is most characteristic of the imperfect setting action of a lime mortar.

The Influence of Calcination upon Cement Action.—The calcination of these varieties of cements plays a very important part in their subsequent behaviour, when tempered with water. Thus it is possible from the same clay-limestone to prepare

(a) A hydraulic **lime**;
(b) A quick-setting cement; **and**
(c) A cement resembling Portland **cement in** character.

At a low temperature in the kiln the mixtures of lime and clay have not mutually reacted the one on the other, and we obtain a material in which the energy due to the hydration of the lime overcomes the tendency of the silicic acid to enter **into combination** with this lime, under the agency of water.

When the second stage in the calcination is reached the silicic acid is liberated or rendered capable of attacking the lime, yielding a cement which sets with comparative rapidity. While, **lastly, under still** more intense firing, the stage of calcination is approached when silicates and aluminates are formed in the kiln and when the material acts like a Portland cement; when the iron, **moreover,** which had during the first and second degrees of **calcination** remained in the condition of a peroxide, passes into that of a protoxide (as is always the case in perfectly prepared Portland cement). This change in the oxide of iron is only effected at very high temperatures, and furnishes a certain indication of the production of a dense slow-setting cement. If in the case of this clayey limestone the clay had been less in quantity we should have obtained a hydraulic lime which would slake with difficulty, and which would be liable to the evil effects of "after-slaking." If the proportion of the bases contained in the clay, as respects the amount of silicic acid present, had been greater, the mass would have probably become vitrified or partially fused before the temperature necessary for the final stage of calcination was reached.

Composition of the Clay Important.—The composition of the clayey matters combined with the lime requires to be carefully studied, as it is found that only under certain conditions does the silica assume the form which causes it to gelatinise when treated with dilute acids.

In this state the silicic acid is spoken of as being soluble, as opposed to the quartz-like reaction of the uncombined silica. Some clays contain a high percentage of sandy particles, or of nearly pure silica not in combination with lime, iron, or alumina, and these clays, though useful to the brickmaker, are ill-adapted for cement making. They are generally characterised by a harsh gritty touch, when tested between the finger and thumb, and it is possible to wash out a considerable percentage of sandy particles.

Characteristics of Clays adapted for Cement Making.— The best clays for the cement manufacturer are those having a greasy unctuous feeling, quite smooth to the touch.

As a rule clays which stain the fingers should be avoided, as being either too much impregnated with iron compounds, or containing a large proportion of organic or other impurities. This does not hold good in the case of the carboniferous shales, some of which are rich in matters which assist in the calcination of the cement. Shales which contain much alum, selenite, or iron pyrites, and many of the shales having a high percentage of carbonate of lime, need great care in manipulation, as they are apt to fluctuate widely in composition and to lead to mistakes in the proportions of the ingredients. We shall speak of shales of this kind when describing the manufacture of Portland cement from the lias formation.

The Influence of the Alkalies.—We have still to deal with the alkalies and their combinations as affecting cement action in conjunction with lime. The proportion of alkaline matters present in the clay used for cement varies, as we shall see later, very considerably ; but the presence of the alkalies is of the utmost importance to the cement manufacturer, for many of the chemical changes upon which he relies are largely assisted by the formation of alkaline compounds. It must be remembered that the silicates of potash and soda are the only ones that are soluble, and the alkalies therefore act as carriers of the silicic acid to the lime and the alumina in the presence of water. Moreover, it is, as already stated, a well-known chemical fact that silica when acting as an acid has a great tendency to enter into combination with a group of bases rather than with any single substance. It is probably for this reason that compound silicates of lime, alumina, iron, and the alkalies are so readily formed in the kiln.

Fremy's Experiments.—From the experiments of M. Fremy, to which we shall refer at considerable length when we consider in detail the kiln changes during the calcination process, it appears that lime and silica react upon each other at a much lower temperature than lime and alumina; silicates of lime being produced at temperatures below 700° Centigrade (1,292° Fahrenheit); while the alumina and lime or the silicate of lime and alumina only begin to form double compounds when that temperature is exceeded. The heat of the kiln at which incipient fusion takes place is about 1,600° Centigrade (2,912° Fahrenheit). It seems, however, probable that alkaline matters have the property to influence the reaction of the lime, alumina, and silica, as also that of the iron, in the cement-material at much lower temperatures, acting in the nature of fluxes, and we think that the knowledge derived from experiments with lime and silica, or lime, silica, and alumina in the absence of the alkalies, is liable to be misleading. The alkalies, after they have done their work during the earlier stages of the calcination, being relatively volatile, are to a considerable extent driven off when extreme temperatures are reached, and they are found only in small quantities in well burned cements. We shall treat of this question hereafter when discussing the induration of cements.

CHAPTER II.

RETROSPECTIVE AND HISTORICAL REVIEW OF THE CEMENT INDUSTRY.

Smeaton's Discoveries.—As we have stated in our opening chapter, the investigations of Smeaton undoubtedly paved the way for all the modern improvements in limes and cements, and led to the overthrow of theories and prejudices which were as old as civilisation itself.

The discoveries of Smeaton, though they took place about 1756, were not published till many years later. In fact it was not until 1791 that he described his experiments, in the third chapter of Book IV. of his *Narrative of the Building, &c., of the Eddystone Lighthouse.*

Smeaton tells us that having been taught a simple method of analysis by Mr. Cookworthy (to whom, by-the-bye, we owe the first use of kaolin in this country, which he discovered in Cornwall and with which he subsequently produced the highly-prized porcelain of Plymouth and Bristol), he tested many varieties of lime, and soon found that the "acquisition of hardness under water did not depend upon the hardness of the stone (from which the lime was made), inasmuch as chalk lime appeared to be as good as that burnt from Plymouth marble, and that Aberthaw lime was greatly superior to either for aquatic buildings, though scarcely so hard as Plymouth marble."

Smeaton's Practical Experiments.—Till this point Smeaton had been guided by the experience of the past, for all writers on building, from Vitruvius to Belidor, had maintained this theory—viz., that lime burnt from dense hard stone made the hardest mortar. But Smeaton's practical tests upset all these traditions, and he sought for some better explanation. He goes on to say "I was very desirous to get some light into some of the sensible qualities that might probably occasion the difference, or at least be a mark of distinction." He then proceeds as follows :—

"Perhaps nothing will better show that the qualities of lime for water mortar do not depend on hardness or colour than a comparison of the white lias of Somersetshire (which, though approaching to a flinty hardness, has **yet a** chalky appearance) with what is called near Lewes in Sussex **the** clunch lime; a kind **of** lime in **great repute there** for **water** works, and indeed deservedly **so. This is no** other than **a** species **of** chalk, not **found** like **the lias in thin** strata, but in **thick masses** as **chalk generally is ;** it is considerably **harder** than common chalk, **but yet of the lowest** degree **of what may** be denominated **a stony hardness ;** it is heavier than common chalk **and not near so white, inclining** towards a yellowish ash colour. **This stone when analysed is** found to contain $\frac{3}{16}$ part **of** its **weight of yellowish clay, with a** small quantity of sand seemingly **of the** crystal kind, **not** quite transparent, but intermixed with red spots."

Hydraulicity Depends on Presence of Clay.—"Hence the fitness of lime for water building seems neither to depend upon the hardness of the **stone,** the thickness of the stratum, nor the bed or matrix **in which it is** found, nor merely on the quantity **of** clay it contains, but, in burning and falling down into a powder of a buff-coloured tinge, *and in containing a considerable quantity of clay,* I have found all the water limes to agree. Of this kind I esteem the lime from Dorking in Surrey to be ; **which** is brought to London under the idea of its being burnt from **a** stone,* and in consequence of that, of its being stronger than the chalk lime in common use there ; though in fact it is a chalk, and not much harder **than** common chalk, it contains $\frac{1}{11}$ part of light coloured **clay of a** yellowish tinge."

Colour not of itself a sufficient indication.—"There is in Lancashire a lime famous for water building called Sutton lime ; I have lately had an opportunity by favour of John Gilbert, Esq., to get a specimen of the stone in its natural state. I had long since seen it in the Duke of Bridgewater's works, both in the burnt stone and slaked, made up for use and in the water. I observed that it agreed with the lias in being of the buff cast. The stone itself is of a deep brown colour, and the piece I have is from a stratum about three inches thick, with a white clayey coat on each side. The goodness of the quality as water lime does not, therefore, consist in the colour before it is burnt ; for we have already seen blue, whitish, and now brown, to be all good for that purpose, but they all agree in the colour or hue after they are burnt and quenched: and having analysed the Sutton limestone I find it to contain not only near $\frac{3}{16}$ part of the original weight **of the** stone, of brown **or red** clay, but also $\frac{1}{12}$ of **fine** brown sand, so that in reality I have seen no lime yet, proved to be good for water building, but what on examination of the stone

* See **Appendix** H, Dorking Stone Lime, p. ~~122~~.

contained clay : and though I am very far from laying this down as an absolute criterion, *yet I have never found any limestone containing clay in a considerable quantity but what was good for water building*, and limes of this kind all agree in one more property, that of being of a dead frosted surface on breaking, without much appearance of shining particles."

Use of Trass and Puzzuolana.—In the same spirit of painstaking research Smeaton experimented with mixtures of lime with trass and puzzuolana, which substances were known at that day to impart to fat limes hydraulic properties. For reasons, which he explains, he rejected the former and employed a mixture of blue lias lime from Watchet and puzzuolana in equal parts, as this was, he conceived, the best water cement that could possibly be made. He likewise tried many other substances mixed with lime, such as the scales from a smith's forge and calcined iron ore.

The lime of Watchet, which he had found to be identical with that of Aberthaw (the beds being merely intersected by the estuary of the Severn), he caused to be slaked on the spot with just sufficient water to enable it to fall down to a fine powder, and this slaked lime was then packed in water-tight casks and sent to Plymouth, whence it was conveyed to the Eddystone rock for use as required.

The Varieties of Mortars used by Smeaton.—Smeaton gives an interesting table showing the composition of the various kinds of mortar he employed, as also the cost of each :—

No.	Name of the mortar.	Winchester bushels by striken measure of			No. of cubic feet.	Expense per cubic foot.
		Slaked lime powder.	Puzzuolana.	Common sand.		
						s. d.
1	Eddystone mortar,	2	2	...	2·32	3 8
2	Stone mortar,	2	1	1	2·68	2 1½
3	,, 2nd sort,	2	1	2	3·37	1 7½
4	Face mortar,	2	1	3	4·67	1 4
5	,, 2nd sort,	2	½	3	4·17	1 1
6	Backing mortar,	2	¼	3	4·04	0 11

Smeaton is said by General Pasley * in his well-known work, to which we shall often refer, to have been the first English engineer to use puzzuolana, which he read of in the works of

* *Observations on Limes, Calcareous Cements, Mortars, Stuccos, and Concrete.* . . . By C. W. Pasley, C.B. London, 1838. 8vo.

Belidor, but for want of chemical knowledge Smeaton failed to appreciate the full significance of the facts he had observed.

Parker's Patent in 1791.—Within a very few years of the publication of Smeaton's work we find the record of Parker's invention of the cement made from " certain stones or argillaceous productions." His name, indeed, first appears in the " Patent Library Index " in 1791, when James Parker, of Christ Church, in the County of Surrey, claims, in Patent No. 1,806, the sole right to employ " a certain material never before made use of" for " burning bricks and tiles, and calcining chalk, earth, stone, and limestone." This he discloses in his specification to be " peat or bog," which was to be interstratified in the kiln with the material to be calcined. As an indication of the quantity needed, he states : " as, for instance, if it requires 36 bushels of coal to burn 220 bushels of lime, the same may be burnt or calcined with 220 bushels of peat with the addition of 3 bushels of coals."

Parker's Second Patent for Cement.—We cannot say with certainty, though it seems probable, that this was the same James Parker who on June 28th, 1796, claims, in Patent No. 2,120, the invention of " A certain Cement or Terras [trass] to be Used in Aquatic and other Buildings, and Stucco Work." He describes his process as follows :—" The principle and nature of the said Invention consists in reducing to powder certain stones or argillaceous productions, called noddles (*sic*) of clay, and using that powder with water so as to form a water (mortar ?) or cement stronger and harder than any mortar or cement now prepared by artificial means. I do not know of any precise generical term for these noddles of clay ; but I mean by them certain stones of clay or concretions of clay, containing veins of calcareous matter, having frequently but not always water in the centre, the cavity of which is covered with small crystals of the above calcareous matter, and the noddles agreeing very nearly in colour with the colour of the bed of clay in or near which they are found. These noddles, on being burnt with a heat stronger than that used for burning lime, generally assume a brown appearance and are a little softened, and when so burnt and softened become warm (but do not slack) by having water thrown upon them, and being reduced to powder after burning, and being mixed with water just sufficient to make them into a paste, become indurated in water in the space of an hour, or thereabouts. Any argillaceous stone, then, corresponding with this description, whether known by the name of noddles of clay, or any other name, is the sort and kind only that I mean to appropriate to my own use in the fermentation (formation ?) of my cement."

The Preparation of the Cement.—"The manner in which I prepare and compose this cement is as follows, viz.:—The stones of clay or noddles of clay are first broken into small fragments, then burnt in a kiln or furnace (as lime is commonly burnt) with a heat nearly sufficient to vitrify them, then reduced to a powder by any mechanical or other operation, and the powder so obtained is the basis of the cement."

Mode of Using the Cement.—"To compose the cement in the best and most advantageous manner, I take two measures of water and five measures of the powder thus described; then I add the powder to the water, or the water to the powder, taking care to stir and beat them during the whole time of intermixture; the cement is then made and will set or become indurated either in ten or twenty minutes after the operation has ceased, either in or out of water.

"But although I have described what I consider as the best proportions for the composition of the cement, it is expressly to be understood that these and all other proportions are to be included within the meaning and purpose of this specification, but that no other proportions will produce so strong a cement in so short a time as those I have here pointed out; and also that I occasionally burn, and grind, and mix the powder before described with lime and stones, clay, sand, or calcined earth in such proportions as may be necessary and useful for the purpose that the cement is intended to be applied to, always observing, the less water is used the better, and, the sooner the mortar or cement is used after being made, the stronger and the more durable it will be."

Parker's Cement termed Roman Cement.—In the above patent Parker is described as of "Northfleet, in the County of Kent, gentleman." According to Pasley it was not until some years later that he applied to the new material the misleading name of "Roman" cement. He shortly afterwards traded in partnership with Mr. Wyatt, and used the stones found at Sheppey in the manufacture of the cement, which soon became widely known under the name of "Parker's cement." A Mr. Samuel Wyatt had previously acquired experience in the use of materials for stucco, and his name is frequently mentioned by Dr. Higgins in his work on Cements.*

Derivation of the Septaria.—The pebbles used for cement making were obtained at various places on the Kentish coast; those from Minster Manor and Whitstable being apparently in

---

* *Experiments and Observations made with the view of improving the art of Composing and applying Calcareous Cements,* . . . by Brindley Higgins, M.D. London, 1780. 8vo.

the best repute. It was not until very much later that Frost
made use of the Harwich-dredged stone, after which time it
was customary to employ a mixture of Harwich and Sheppey
stones in equal proportions to burn the best Roman cement,
though Pasley tells us that about 1836 in the Government
dockyard at Sheerness they used one part of Sheppey to three
parts of Harwich cement stone. The makers of artificial cements
about this period sold the cement made from natural stone at a
much higher price than that made from a mixture of chalk and
clay.

Pasley's **Tests** of Cements in 1836.—The following table
compiled from Pasley's book is valuable as showing the comparative merits of the chief cements in use at the time he wrote
[1836]. All the specimens were tested neat :—

Sort of Cement used.	Age in Days.	Total Tensile Strength in Pounds.	Lbs. per Sq. Inch.
Pasley's artificial cement,	11	1,395	34·9
Frost's ,, ,,	11	705	17·6
Francis (Natural), . .	11	1,223	30·6
Sheerness ,, . . .	{ 10 } { 12 }	1,220	30·5

French Quick-Setting Cements.—The quick-setting cements
of the Roman type, made in accordance with Parker's patent,
had their counterpart in France, for about 1796 a French military engineer, named Lesage, drew attention to the eminently
hydraulic properties of the pebbles of Boulogne-sur-Mer, and
from these he produced a material which he called "plaster-
cement," though this substance was quite distinct from the
compounds of the calcic sulphate series.

Another quick-setting cement, which obtained a wide reputation on the Continent, was that of Pouilly, discovered by Mons.
Lacordaire in 1829. Four years later, in 1831, Mons. H. Gariel
first introduced the well-known cement of Vassy. Another
cement of a similar character, which has been much used in
France, is made from stone dug near Grenoble at Porte-de-
France.

Analyses of French and English Cements.—The composition of all these materials is very similar, as will be seen by
the following set of analyses, made in Paris, at the École des
Mines :—

	English Roman Cement.	Boulogne Cement.	Cement of Vassy.	Cement of Pouilly.
Silica, ...	15·42	14·27	16·0	17·14
Alumina, . . .	5·63	5·81	6·50	6·87
Iron, . . .	6·81	5·97	8·60	4·83
Magnesia, . . .	0·83	0·67	1·50	3·39
Manganese, . .	0·54	0·83	...	0·17
Lime, ...	45·12	46·20	46·80	47·35
Loss on calcination ,mois- ture, and carbonic acid, }	25·65	26·25	20·60	20·25
	100·00	100·00	100·00	100·00

The Vassy cement, which is very dark in colour, is the richest in iron, and sets when freshly burned with great rapidity. The cement of Grenoble is very similar in its composition to the above, but contains rather less clay. In France the quick-setting cements are, as a rule, ground under edge-runners and passed through copper wire sieves of 324 meshes per square centimetre (2,090 per square inch). The average specific gravity of these cements varies from 0·96 to 1·10.

Rosendale Cements of America.—The American magnesian cements likewise belong to this category. Among the best known of these quick-setting cements, all of which incline rather to the type of what are known in this country as Roman cements, is that named after the town of Rosendale, where it was originally discovered and largely used in the construction of the old Delaware and Hudson Canal.

Raw Materials for American Cement.—The formation which furnishes the raw materials for this cement is an argillaceous magnesian limestone, which extends along the whole of the Appalachian range, and is said to yield three-fourths of the hydraulic cement produced in the United States. The workable beds are seventeen in number, and the method of manufacture is as follows :—

The limestone, which is blasted by black powder, is conveyed to the kilns on trucks running along the base of the quarry. The stone is arranged in the kilns in layers interstratified with small coal. The kilns are cylindrical in form, but have a conical contraction at the bottom. They are worked continuously, portions of the charge being withdrawn every twelve hours. The underburnt lumps are passed a second time through the kiln ; the fully calcined stone is conveyed in elevators to the top of the mill. Here it is first broken to a small gauge (size of hazelnut) in a cracker, which is a machine somewhat on the principle of the coffee-mill. Each cracker will prepare from 250 to 300 barrels of cement *per diem* for the mills.

Grinding of the Calcined Cement.—The mill-stones are only 3 feet in diameter, but they grind the cement so fine that from 93 to 95 per cent. will pass a 50 × 50 mesh sieve. These mills differ from those in use in this country for cement-grinding, in that what is termed in England the "runner" or the upper stone is fixed and the lower one revolves. The ground cement is conveyed by "creepers" to the packing room, where it is put up in paper-lined casks and is then ready for use.

Yield and Composition of the Beds.—Each cubic yard of cement stone yields about 2,700 lbs. of finished cement. The various beds of stone fluctuate very considerably in their composition, and the preparation of the cement and the mixture of the raw materials from the different levels is a matter of great nicety. During the manufacture the cement is carefully tested; four briquettes being made every hour. The pattern used for the moulds differs from that in common use in Europe.

General Gillmore's Tests.—General Gillmore, whose work on cements * may be consulted with advantage respecting the manufacture and employment of these materials, puts the crushing strength of the neat cement at 546 lbs. per square inch in seven days, and 2,015 lbs. per square inch in thirty days. The tensile strength of the neat cement is 104 lbs. per square inch in seven days, and 134 lbs. per square inch in thirty days. With an equal amount of sand the tensile strength in thirty days is 102 lbs. per square inch.

Early Cement Patents in England.—The records of the Patent Office of this country show us that prior to the advent of Parker some eighteen inventors had busied themselves with processes for burning lime, for the manufacture of "tarrass" or puzzuolana, for a composition or cement called "Pietra Cotta," and for sundry descriptions of mastic, mortar, and stucco. Among these we find the "water cement or stucco" described in the work already mentioned of Dr. B. Higgins. This patent is No. 1,207, and is dated Jan. 8, 1777.

From the date of Parker's second patent in 1796 to that of James Frost's specification in 1822, the Patent Index likewise includes eighteen patents, one of the most notable of which was that granted to St. Leger in 1818.

St. Leger's Patent in 1818.—St. Leger was probably the concessionaire of Vicat, a French engineer, whose experiments

* *Practical Treatise on Limes, Hydraulic Cements, and Mortars*, by Q. A. Gillmore, A.M., New York, 1874.

with limes and cements are justly famous, and his work may
here receive passing notice. Omitting the investigations of
Loriot, published in 1774, of De La Faye, said to have been
made known in 1777, and the artificial stone of Fleuret, described in his book dated 1807, all of which are duly recorded
by Pasley, we come to the publication of Vicat in 1818.* It
will be impossible here to do more than refer most briefly to his
treatise, but as his later work in 1828† has been ably trans-
lated by Captain Smith, and as General Pasley alludes at considerable length to his investigations in the appendix to his
"Observations on Lime, Calcareous Cements, &c.," we need not
dwell on them very fully.

Vicat's Artificial Lime.—Vicat, who appears to have been
the first to apply the term "hydraulic" to limes for use in water,
prepared mixtures of pure lime and clay, and calcined the resulting material in kilns. He employed, originally, slaked lime powder
together with 20 per cent. of clay, but in his second treatise
he describes a plan of grinding chalk along with the requisite
proportion of clay. His process was actually carried out at
Meudon by Messrs. Bryan and St. Leger, who used the chalk
found in the vicinity, and the plastic clay of Vaugirard. They
ground together in a washmill four measures of chalk and one of
clay with a considerable quantity of water, and the fluid slip
was then run into reservoirs to dry. When stiff enough the
pasty mass was moulded into blocks and burnt in kilns, interstratified with small coal and coke. This plan of making an
artificial lime was still practised at Meudon when the author
visited the works in 1878.

Vicat shares with Dr. John of Berlin the honour of having
explained the importance of the presence of clay in combination
with lime on scientific grounds, though De Saussure first pointed
out that hydraulicity depended solely on the presence of clay,
and Descotils in 1813 observed that this phenomenon was due
to the existence of a large quantity of siliceous matter dis-
seminated through the mass in very fine particles. Dr. John's
treatise was published in 1818. Smeaton, as we have stated,
anticipated by many years these discoveries, though he failed to
turn them to practical account.

Specification of St. Leger's Patent.—Doubtless it was a
process similar to the above which was, as we have seen, patented

* Recherches Expérimentales sur les Chaux de Construction, les Bétons et
les Mortiers Ordinaires, par L. J. Vicat. Paris, 1818. 4to.
† Résumé des connaissances positives actuelles sur les qualités, le choix et la
convenance réciproque des matériaux propres à la fabrication des Mortiers et
Ciments Calcaires, &c., par L. J. Vicat. Paris, 1828.

in this country in 1818 (No. 4,262). M. Maurice St. Leger describes himself in the specification as of St. Giles, Camberwell, in the County of Surrey. The patent is for "an improved method of making lime" out of a fat or pure carbonate of lime by the addition to the same of "clay or any substance containing alumine and silex." The general proportion named by St. Leger is "from 1 to 20 measures, or given quantities of clay or other substance containing alumine and silex, to every 100 measures or given quantities of chalk, stone, or other substance, or of lime." It does not appear that St. Leger worked the patent in England.

Frost's Patent for British Cement in 1822.—On the 11th of June, 1822 (No. 4,679), James Frost, of Finchley, in the County of Middlesex, Builder, specifies "a new cement or artificial stone," which is made as follows:—"I select such limestones or marls, or magnesian limestones or marls as are entirely or nearly free from any mixture of alumina or argillaceous earth, and contain from 9 to 40 per cent. of silicious earth or silica, or combinations of silica and oxides of iron, the silica being in excess and in a finely divided state, and break such selected materials into small pieces, which are then calcined in a kiln, in the manner calcareous substances usually are, until all carbonic acid be expelled, and until it be found on trial of a small portion of such calcined materials that it will not, when cool, slack or fall when wetted with water. The calcined material is to be ground to a fine powder by any machinery fit for reducing dry substances to that state, and the powder is the material for making the cement or artificial stone, and must be kept in dry packages for use. When used it is to be mixed with water, and tempered to the consistency of common mortar; it should be mixed in small quantities and applied instantly to its intended purpose, as it will set in a few minutes to resist the impression of the finger and gradually harden to a stony body. For many purposes a quantity of clean silicious sand may be advantageously incorporated with it when it is tempered for use. The cement will be lighter or darker in colour, as there is a lesser or greater quantity of oxide of iron in the selected materials; the lighter colour will be found best adapted to dry, and the darker colour to wet situations.

"I declare that my invention is for making a cement or artificial stone from siliceous limestones or marls in the manner hereinbefore described, and that I shall call it by the name of British cement."

There are some singular blunders in this specification which prove conclusively that Frost had a poor grasp of the subject.

Thus he proposes to use limestone or marl "free from any mixture of alumina," but containing "from 9 to 40 per cent. of silica," which would be pure sand, and would inevitably yield a compound devoid of cementitious properties.

Frost's Second Patent.—In 1823 (No. 4,772), Mr. Frost describes a process of "calcining and preparing calcarious (*sic*) and other substances for the purpose of forming cements," with which object he used a kiln of special construction, in which the material was permitted to cool very slowly; this is an invention of no importance, and it was probably founded on a misconception.

Frost as a Cement Maker.—Very shortly afterwards, in 1825, Frost established himself in cement works at Swanscombe, on the Thames, and became the first maker of artificial cement in the London district. Frost's business was carried on by him until 1832 or 1833, when his interest was acquired by Mr. John Bazley White, Senr., who at first was the partner of Mr. Francis. Pasley tells us that Messrs. Francis, White, & Francis "were in the habit of selling their artificial cement at one shilling per bushel at the same time that they sold their natural cement for eighteenpence."

Mr. Frost, when he had disposed of his business, retired to America, and we can find no further records of his career. His works were carried on by Messrs. Francis & White in partnership until January 1st, 1837, when they were made over entirely to Messrs. White & Son, who had also works at Millbank. Mr. Francis, who traded as Messrs. Francis & Sons, founded the well-known works at Nine Elms, Vauxhall.

In strict historical sequence we should before this have referred to Aspdin and his invention of Portland cement, but this may form the subject of a special chapter.

CHAPTER III.

THE EARLY DAYS OF PORTLAND CEMENT.

Portland Cement first made by Aspdin.—The somewhat misleading name of "Portland" cement was given to the artificial compound of lime and clay, prepared by Mr. Joseph Aspdin, a Leeds bricklayer. He chose this name in consequence of its fancied resemblance in point of colour and texture to the oolitic limestone of the island of Portland, well known and in great favour in this country as a building stone.

Aspdin's Specification.—Aspdin's specification, No 5,022, is dated October 21st, 1824, and is for "An Improvement in the Modes of Producing an Artificial Stone," which invention he thus describes :—" My method of **making a cement or artificial stone** for stuccoing buildings, **water** works, **cisterns, or any other** purpose to which it may be applicable (and which **I call Portland** cement) is as follows :—I take a specific quantity **of limestone,** such as that generally used for making or repairing **roads, after** it is reduced to a puddle or powder ; but if I cannot procure a sufficient quantity **of the above from the** roads, I obtain **the** limestone itself and I cause **the puddle or** powder, or the lime-**stone as** the case may **be, to be calcined.** I then take a specific quantity of argillaceous earth **or clay,** and mix them with water to a state approaching impalpability, either **by** manuel (*sic*) labour or machinery. After this proceeding I put the above **mixture** into a slip pan for evaporation, either by the heat of **the sun or** by submitting **it to** the action of fire **or** steam conveyed in **flues or** pipes **under or near the pan, until** the **water** is entirely evaporated. **Then I break the said mixture into** suitable lumps, **and calcine them in** a furnace similar to a lime kiln till the **carbonic acid** is entirely expelled. The mixture so calcined is **to be ground, beat,** or rolled to a fine powder, and is then in a fit **state for making cement or** artificial stone. This powder **is to be mixed with a** sufficient quantity **of water** to bring it into the consistency of mortar, and thus **applied to** the **purposes wanted."**

Aspdin's Second Patent.—Aspdin, whose name, by-the-bye, in the "Alphabetical Index of Patentees" is misspelled "Apsdin," obtained **protection in the following year,** 1825, under date of

June 7th (No. 5,180), for "a Method of Making Lime." In order to effect this, he collected the road scrapings from roads repaired with limestone; these he dried, either by natural or artificial heat, and they were subsequently "removed to a furnace or kiln to burn with coal, coke, or wood." The product was then ready either for "building, or liming land."

Aspdin's Process differs from that used subsequently.— It is difficult to recognise in this description a process likely to result in the formation of a cement of the present Portland type. It must be remembered, however, that Aspdin had a hard mountain limestone to deal with, and that probably the most easy way to obtain this material in a state of fine subdivision, in order to mix it with the clay, was to calcine it. It could then readily be slaked and reduced to powder. The next step was to temper it with the requisite amount of clay, and finally the mixture was submitted to a second process of calcination. This double-kilning would, where fuel was relatively cheap, entail but little more cost and perhaps less labour than first grinding the limestone to fine powder under mill-stones and then mixing it with the clay, as is now done in the dry process of manufacturing Portland. Moreover, by the slaking action, the lime is obtained in an extremely fine state of subdivision, and therefore in a condition peculiarly well adapted for intimate admixture with the clay.

Defects in Aspdin's Specification.—Aspdin fails to point out the exact amount of clay needed, rather an important matter in a specification, one would think, and he omits to state that the firing must be carried on until incipient vitrification is attained; these omissions might by some be regarded as casting doubts upon the authenticity of his discovery of Portland in 1824, but it is a well-known fact that he had a manufactory of Portland for many years at Wakefield, which was established in 1825, and which is still in existence; moreover, his son, William Aspdin, was, as we shall see, one of the early cement manufacturers in the Thames district.

Proposals to make Cements from Artificial Mixtures.— On searching the Patent Office records it is found (as we have already mentioned) that other inventors had attempted to produce cements from mixtures of lime and clay, and Pasley has, in his important work on cements, previously noticed, recorded a long series of patient investigations and experiments, having for their object the formation of an artificial cement capable of giving like results to those obtained from the natural cements, made from the septaria of the London clay, and the nodules of argillaceous limestones found in the Isle of Sheppey.

Roman and Quick-setting Cements.—In our previous

chapter we have given a brief account of these materials, and of the artificial hydraulic lime of Meudon, the invention of Vicat. Cements of this type were in the early years of the present century in very general use in this country, and many attempts were made by inventors to obtain a cheaper source of supply. It was clearly the search after such substitutes for the natural cement stone that led to the discovery of Portland cement, and though this latter material has now very largely displaced the quick-setting cements of former days, there are no doubt many purposes for which Roman cement can be used with advantage, and this cement is still in constant demand.

Early History of Portland Cement by Pasley.—We know but little of the early history of Portland cement. General Pasley, in the preface of the second edition of his work, dated August, 1847, states : " At present there are three manufactories of artificial cements in England, which have all been used more or less extensively in works of importance, and have given satisfaction, viz.:—First, that of Messrs. John B. White & Sons, in the Parish of Swanscombe, Kent, the present proprietors of Mr. Frost's Works, who, after gradually relinquishing the objectionable parts of his process, have succeeded in making a good artificial cement, which they call their **Portland cement,** by a mixture of chalk found on their own premises with the blue clay of the Medway ; secondly, that of Messrs. Evans & Nicholson, of Manchester, who make an artificial cement, which has been called the **patent** lithic cement, with the very same ingredients, and in the same proportions, nearly, that were used in the author's experiments, but the most important of which is obtained in a roundabout manner from the residual matters or waste of certain chemical works, instead of working with natural substances ; thirdly, that of Mr. Richard Greaves, of Stratford-upon-Avon, who makes a powerful water cement, which he calls blue lias cement, by mixing a proportion of indurated clay or shale with the excellent blue lias lime of that neighbourhood, both of which are found in the same quarries ; the former being previously broken and ground, and the latter burned and slaked, which is absolutely necessary in making an artificial cement from any of the hard limestones."

Pasley's Letter to Dr. Garthe.—This account of the Portland cement industry was, however, incorrect, for we find in a letter published in Dingler's *Polytechnisches Journal,** written by Pasley to Dr. Garthe of Cologne, and dated March 3rd, 1852, the following additional information on the subject :— " I am much flattered by the favourable opinion you express

* Vol. cxxiv., p. 27.

concerning my work on cement. As shortly after my promotion to the rank of Major-General I ceased to occupy the post of Inspector of the Royal Engineers' School at Chatham, and had, therefore, neither assistants, materials, nor appliances at my disposal, I was no longer in the position to prosecute researches of a similar nature to those formerly carried on by me, and which resulted in the discovery of an artificial compound, but little inferior to the best natural cements.

"For several years past I paid no further attention to the subject until I learned that Portland cement was superior to Roman cement ; and as it is the fashion in our country to disguise everything with some fantastic name or other, which, except in the case of chemical products, gives not the slightest clue to the composition of the article in question, I was astonished to discover that this Portland cement, the name of which would lead the foreigners visiting our great Industrial Exhibition of last year to believe very naturally, either that it was a cement found in the island of Portland, or that it was related in some way or other to Portland stone, was neither more nor less than my own artificial cement, compounded of chalk and clay.

"Messrs. Robins, Aspdin & Company manufacture Portland cement which appears to me to be just as good as, if not superior to, that of Messrs. White & Sons, although I never heard of it until I met Mr. Aspdin in the great Exhibition last year. I was present at all the experiments upon Portland cement mentioned in *The Builder* of the 27th September last. The results of the same are correctly reported, but several of them are rendered ambiguous for want of more complete particulars or explanatory drawings relating to them."

Becker's Book on Portland Cement in 1853. — In a German work on Portland cement by Becker,* in which this letter is quoted, and which we believe was the first treatise in any language on this material, there is a footnote calling attention to the singular fact that "although Joseph Aspdin, the inventor and patentee of Portland cement, established his manufactory at Wakefield in 1825, and his son, William Aspdin, founded his works at Northfleet on the Thames about ten miles from Chatham a few years later, at which latter place Sir C. W. Pasley was then residing, and although their Portland cement was already much employed on the Continent, as introduced in the first instance by Maude & Son, and subsequently by Robins, Aspdin & Company, whereas the cement of Messrs. White & Sons was not brought into the market until many years later (in 1845), Sir Charles Pasley only heard for the first time of the existence

* *Erfahrungen über den Portland Cement*, Berlin, 1853. 8vo.

of Joseph Aspdin, the inventor of Portland cement, and of his
manufactory [at Wakefield] in the great Exhibition of 1851."

Aspdin, the Younger.—Mr. Becker is, however, not quite
correct in his facts, for young Aspdin did not go to Northfleet
until 1848. It appears that he at first associated himself with
Messrs. J. M. Maude, Son & Company, of Upper Ordnance
Wharf, Rotherhithe, and from a circular issued by this firm in
1843 we learn the following particulars respecting their business,
and relative to the cement industry at that date :—

Circular by J. M. Maude, Son & Company.—"The manufacture of this cement (patent Portland cement) has for many
years been carried on by Mr. Aspdin at Wakefield, in which
neighbourhood, as throughout the northern counties of England,
it has been successfully and extensively used ; owing to the heavy
charges attending its conveyance to the London market its
consumption there has necessarily been limited, and although
its superiority over other cements has never been contested by
those who have been induced to give it a trial, the high price at
which alone it could be supplied has hitherto proved a serious
impediment to its more general introduction into the metropolis.
Messrs. J. M. Maude, Son & Company have now the satisfaction
of announcing to the public that they have made arrangements
with the son of the patentee for the purpose of carrying on the
manufacture of this valuable cement at their extensive premises
at Rotherhithe ; and whilst they will thus be enabled to
supply it at a considerably reduced price, they have also the
satisfaction of stating that in consequence of improvements
introduced in the manufacture, it will be found, for the following
reasons, infinitely superior to any cement that has hitherto been
offered to the public :—

"(1) Its colour so closely resembles that of the stone from which
it derives its name as scarcely to be distinguishable from it.

"(2) It requires neither painting nor colouring, is not subject
to atmospheric influences, and will not, like other cements,
vegetate, oxydate, or turn green, but will retain its original
colour of Portland stone in all seasons and in all climates.

"(3) It is stronger in its cementitive qualities, harder, more
durable, and will take more sand than any other cement now
used."

After alluding to its value as a stucco and for paving purposes,
the circular states—"It is manufactured of two qualities, and
sold in casks of 5 bushels, each weighing about 4½ cwts.—
No. 1 at 3s. per bushel, or 20s. per cask, and No. 2 at 2s. 3d.
per bushel, or 16s. 3d. per cask. 4s. 6d. each allowed for the
casks, if returned in good condition." From the directions for its

use we learn that the best quality was to be employed with four parts of sand.

Early Tests of Portland Cement.—Certain particulars are likewise given of some comparative trials of this Portland cement with Roman cement, which were conducted by Messrs. Grissell & Peto at the new Houses of Parliament in 1843, together with a letter from these eminent contractors, dated from York Road, Lambeth, November 13th, 1843, setting forth the fact that the experiments in question, which consisted of bricks stuck one on to another projecting from a wall, and of brick beams with bearings of 3 and of 5 feet, were made by their direction and under their own superintendence, and they add—"The results as shown by the above statement afford very satisfactory evidence of the superior qualities of your cement." They conclude that Portland with three parts of sand was more than double the strength of Roman cement with one part of sand, and that with Roman at 1s. 4d. and Portland at 2s. 3d. the bushel there was a saving of from 1½d. to 2d. per bushel of mortar in using Portland, owing to the increased volume of sand this latter cement would carry.

Wylson on Portland Cement.—It was just about this period that Portland began to make a London reputation, and in some articles on "Mortars and Cements" in *The Builder* in 1844, Mr Wylson, speaking of these same experiments, says: "The deductions thence arrived at show an advantage on the side of the Portland cement, which speaks most strongly in its favour; and which, judging by the authority from which these contrasts emanate, must be considered as at once authentic and conclusive, establishing this to be beyond all doubt superior to the Roman cement, whether as to strength, adhesion, or the capacity of receiving sand; the latter of which properties it is shown to possess to such a degree as to render it actually cheaper than the Roman cement, whilst its other recommendations of beauty and the saving of colouring alone render it highly preferable."

Early Advertisements relating to Portland Cement.—Advertisements relating to this cement as made by Messrs. J. B. White & Sons appear for the first time in *The Builder* in October, 1845, and on the occasion of an accident to some stucco on a house in Cornhill, which had failed or "blown" in the following year, Messrs. Maude, Jones, & Aspdin write under date of September 3rd, 1846, to deny that the cement in question was of their manufacture.

Sir William Tite on Portland Cement.—In May of this same year, Mr. W. Tite, the then Vice-President of the Institute of British Architects, spoke of Portland cement as "a good material if properly made and properly applied."

We have now reached the period when the railway fever set in, and the demand for cement stone at Harwich became very extensive, indeed the stock of stone threatened to become exhausted. We read in December, 1846, "prices have in consequence risen 30 per cent., while the dredgers are reaping a proportionate benefit. It is calculated that £25,000 per annum are paid away in wages alone to workmen employed in this trade."

Predicted Exhaustion of Natural Cement **Stone.**— It must have been in the year 1845 or 1846 that Sir R. Peel announced in Parliament his intention of taxing the cement stone, fearing its exhaustion, and hoping thus to reserve a sufficient supply for the purpose of Government works. This fact we learn from a pamphlet issued by Messrs. Aspdin, Ord, & Co., who state also therein that—" As soon as Mr. Wm. Aspdin heard of this he addressed Sir Robert Peel upon the subject, and intimated to him that he need not be alarmed about the supply of 'Roman cement,' as an article, which far surpassed it, was extensively manufactured. To prove these remarks, Mr. Aspdin waited upon Sir R. Peel with samples, and obtained a reply to the effect that he was much satisfied with them, and with the explanation; and in consequence the proposed tax was abandoned."

Portland Cement from Wakefield used in Thames Tunnel.—From this same pamphlet we find that the Wakefield cement was largely used in the Thames Tunnel about 1828, although "at that time it cost 20s. to 22s. per cask, besides the carriage to London. Yet Sir I. Brunel decided (notwithstanding his ability to procure Roman at 12s. per cask delivered on the spot) to adopt Portland chiefly for his purpose, as its merits required no other recommendation than an impartial trial."

From this statement we may assume that the Thames Tunnel was the first engineering work of importance in which Portland cement was used, for there can be no doubt that in its early days it was mainly a stucco cement, and it was not until very much later that it acquired the confidence of the engineering profession.

Accident at Euston **Station** in 1848.—The year 1848 was in many ways a very eventful one in the history of Portland cement. On the last day of January a fatal accident occurred at Euston Station owing to the incautious use of Roman cement. A lofty wall supported on columns formed of brick on edge suddenly fell, causing the death of two workmen, and at the inquest which followed, and which attracted much attention, the question of the employment of cement was discussed at great length. Mr. Hardwick, the architect, expressed an opinion that the failure of the cement was due to its employment during the cold winter weather, and also to the great haste with which the work was run up. The columns in question were 20 feet high, 2 feet 2½ inches in diameter at the base, and 1 foot 10½ inches in diameter below the capital. Messrs W. Cubitt & Company, the builders, stated that they could not assign a reason for the fall of the wall. On February 11th, Mr. William Aspdin wrote from Northfleet and attributed the accident solely to the use of defective cement. We have here an intimation of Aspdin's removal from Rotherhithe.

Experiments at Messrs. Grissell's Works in 1847.— Later on in this same year the results of some trials of Messrs. White's Portland cement are published in *The Builder*. These experiments had taken place at Messrs. Grissell's ironworks on December 10th and 31st, 1847, and they furnished additional evidence of the superiority of Portland over Roman cement.

Aspdin's Circular in 1848.— The above trials and the publicity they obtained seem to have given considerable umbrage to Aspdin and his partners, for in the September following he published a letter as an advertisement in *The Builder*, in which he set forth, among other matters, that this cement was first introduced by his father in 1813, and patented by him as "the Portland cement" in 1824, and that the original material is solely made by him. He quotes certain tests to show that he had obtained far better results than Messrs. White had done, and that he had used more sand. Thus he states that he has made blocks consisting of one part of cement to ten parts of sand, and he concludes by calling attention to a public trial of their cement which the firm were about to institute at their wharf at Great Scotland Yard.

Experiments in 1848 at Messrs. Bramah's Works.— These proposed experiments they advertised for some time, and they challenged all other cement makers to compete with them. The trials came off on the 18th September, and subsequently at Messrs. Bramah's works on the 26th, when some blocks were crushed by hydraulic pressure. We extract the following particulars from *The Builder* of September 30th, 1848, as they show the quality of the cement at that period. The brick tests were of the usual character—viz., built out from the wall. In one case the beam consisted of 38 bricks with neat cement. With one

part sand 15 bricks. Some brick beams were also tested, but
the results are not very clear. Two large blocks of Portland
stone, jointed together with cement, failed in the stone work.
Some blocks of cement, 18 × 9½ × 9, were then tried in a Bramah
press, and stood 58¼ tons, when the press gave way. Roman
cement went under 22½ tons. A block of one part cement
to one sand withstood 45 tons. On the 26th September some
further tests were made with a better press at Messrs. Robin-
son & Sons, Pimlico, when the following results were obtained :—

"1. One of the previous blocks, all cement and then thirty-
five days old, bore a pressure of 68 tons, when it cracked
at one angle, but ultimately bore 90 tons without any alteration.

"2. An exactly similar block bore 141 tons without even a
crack, and stood under that pressure for nearly one minute, when
it broke.

"3. Another similar block bore 104 tons.

"4. One composed of nine sand to one cement bore 4½
tons.

"5. One of equal quantities of sand and cement previously
submitted to a pressure of 47 tons now cracked at that pressure,
but ultimately bore a pressure of 108 tons before it broke down,
the form of the crack not altering."

Public Attention begins to be directed to Portland
Cement.—The rivalry between the two principal firms engaged
in the manufacture of Portland cement, in what we may term
the London district, had the effect of directing public attention
to the new material, and the cement industry from this time
steadily increased in importance. Fresh factories were started,
and the quality gradually became more and more reliable.

Early Experiments on the Strength of Cements.—In a
paper read by Mr. G. F. White before the Institution of Civil
Engineers in May, 1852, there is an account of the experiments,
to which reference has already been made, at Messrs. Grissell's
works in 1847, and of some additional tests carried out at Mr.
Jackson's works on March 8th, 1851. We quote certain of these
results, as they are of importance as illustrating the relative
merits of the artificial and natural cements made about that
time. Each block was 9 inches square and 18 inches long, and
it was tested by hydraulic pressure applied on the smaller ends,
a surface of 81 square inches :—

Nature of Block.	Age in Days	Crushing Force in Tons.		Lbs. per sq. inch.
		On 81 sq. ins.	On 144 sq. ins.	
1. Pure Portland cement, ...	30	75	133	2,074
2. { 1 Portland, } Immersed in water for } 2 Sand, 7 days after making, }	52	45	80	1,244
3. { 1 Portland cement, } Immersed in water 3 Sand, as soon as made for 7 days, }	52	25	44	691
4. Pure Roman cement, ...	30	27	48	746
5. { 1 Roman cement, } 2 Sand, }	52	3	5·33	83
6.* { 1 Portland, } 2 Sand, } Concrete, }	48	24	42·6	664
7.* Do., ...	270	63·75	113·33	1,763
8.* { 1 Portland, } 3 Sand, }	70	16·875	30	442
9.* { 1 Portland, } 10 Shingle, }	30	10	17·77	276

Experiments made at Mr. Jackson's works on blocks 6 inches by 6 inches, and 12 inches long :—

Nature of Block.	Age in Days.	Crushing Force in Tons.		Lbs. per sq. inch.
		On area of 36 inches.	On the sq. foot.	
Pure Portland cement, . . .	40	40	160	2,453
„ Atkinson's „ . . .	40	20	80	1,244
„ Sheppey „ . . .	40	19·5	78	1,213
„ Roman „ . . .	40	13·25	53	829

It is argued from the first set of experiments that neat Portland cement is three times the strength of neat Roman cement, and that with two parts of sand Portland is half as strong again as neat Roman cement, while even with three parts of sand it is equal in strength to neat Roman cement.

* Experimental blocks 6, 7, 8, 9, described as concrete, of which cement and sand occupied one-fifth part in bulk.

3

Experiments at the Exhibition of 1851.—Some experiments tried at the great Exhibition in Hyde Park, before the Jury of Class XXVII. in September, 1851, are interesting, as the cement was used in the briquette form (see Fig. 1).

Fig. 1.

The breaking weight per square inch of neat Portland was 414 lbs., which is a very good result for that period. No indication is, however, given of the age of this sample.

Prices of Roman and Portland Cement.—The price of Roman cement had fallen about 1848 to 1s. 3d. per bushel, the original price having been 4s. 6d., the cost of Portland cement at this time was, as we have seen, about 2s. 6d. per bushel.

Mr. Johnson's Statement respecting the Manufacture.—We think that we may here insert a very important statement respecting the early history of Portland cement, drawn up for the author by the veteran cement manufacturer, Mr. I. C. Johnson, in 1880, extracts from which were published in *The Building News* of that year. Mr. Johnson states that young Aspdin "began work at Rotherhithe in connection with Messrs. Maude & Son on a small scale, and did sometimes make a strong cement, but, owing to want of scientific method, the quality as respects strength and durability was not to be depended upon." He proceeds :—

"I was at this time [about 1845] manager of the works of Messrs. White, at Swanscombe, making only the Roman cement, Keene's plaster, and Frost's cement, the latter composed of 2 chalk to 1 of Medway clay, calcined lightly, and weighing 70 to 80 lbs. per bushel.

"My employers, attracted by the flourish of trumpets that was then being made about the new cement, desired to be makers of it, and some steps were taken to join Aspdin in the enterprise, but no agreement could be come to, especially as I advised my employers to leave the matter to me, fully believing that I could work it out.

"As I before said, there were no sources of information to assist me, for although Aspdin had works, there was no possibility of finding out what he was doing, because the place was closely built in, with walls some 20 feet high, and with no way into the works, excepting through the office.

"I am free to confess that if I could have got a clue in that direction I should have taken advantage of such an opportunity, but as I have since learned, and that from one of his later partners, that the process was so mystified that anyone might get on the wrong scent—for even the workmen knew nothing, considering that the virtue consisted in something Aspdin did with his own hands."

Aspdin's Secret Processes.—"Thus he had a kind of tray with several compartments, and in these he had powdered sulphate of copper, powdered limestone, and some other matters. When a layer of washed and dried slurry and the coke had been put into the kiln, he would go in and scatter some handfuls of these powders from time to time as the loading proceeded, so the whole thing was surrounded by mystery."

Analysis of Aspdin's Cement.—"What then did I do? I obtained some of the cement that was in common use, and, although I had paid some attention to chemistry, I would not trust myself to analyse it, but I took it to the most celebrated analyst of that day in London, and spent some two days with him. What do you think was the principal element according to him? Sixty per cent. of phosphate of lime! All right, thought I, I have it now. I laid all the neighbouring butchers under contribution for bones, calcined them in the open air, creating a terrible nuisance by the smell, and made no end of mixtures with clay and other matters contained in the analysis, in different proportions and burnt to different degrees, and all without any good result."

Attempt to produce a Cement from the Components of Roman Cement.—"The question was, what was the next thing to be done? I had an idea that the elements were those contained in Roman cement, and I had read somewhere that the older chemists had taught that the value of Roman cements was due to the iron and manganese contained in them. I knew that these matters gave rise to the peculiar colour of Roman cement, but they were absent in Portland."

Mr. Johnson's Experiments.—"I had a laboratory and appliances on the premises, so I worked night and day to find out the component parts of the stones from Harwich and Sheppey. Having found these, and having tried many experiments, spreading over some months, in putting different matters together, I began to think that lime and alumina were the chief ingredients necessary. I, therefore, tried quicklime powdered and mixed with clay and calcined, by which means I got something nearer. It was a cement very much like Frost's. After this I used chalk and clay as used in Frost's cement, but with more chalk in proportion. The resulting compound being highly burned, swelled, and cracked.

"By mere accident, however, some of the burned stuff was

clinkered, and, as I thought, useless, for I had heard Colonel
Pasley say that he considered an artificial cement should feel
quite warm after gauging, on putting your hand on it, and that
in his experiments at Chatham he threw away all clinkers
formed in the burning."

Trial of Clinker formerly rejected.—" However, I pulverised some of the clinker and gauged it. It did not seem
as though it would harden at all, and no warmth was produced.
I then made mixtures of the powdered clinker, and powdered
lightly-burned stuff,* this did set and soon became hard. On
examining some days later the clinker only, I found it much
harder than the mixture, moreover, the colour was of a nice
grey."

Works conducted on a Larger Scale.—" Supposing that
I had nearly got hold of the right clue, I proceeded to operate
on a larger scale, making my mixture of 5 of white chalk to
1 of Medway clay. This was well burned in considerable
quantities and was ground finely, but it was of course a failure
from excess of lime, although I did not then know the reason of
it. The whole of this material was tossed away as useless into
a kind of tunnel near at hand, and laid there for some months,
after which I had the curiosity to take a sample of it and gauged
it as before, when, to my astonishment, it gauged smoothly and
pleasantly, and did not crack and blow as before, but became
solid and increased in hardness with time."

Traces a Previous Failure to Excess of Lime.—" Cogitat-
ing as to the cause of this difference, it occurred to me that
there had been an excess of lime, and that this exposure in a
rather damp place had caused the lime to slake.

" This was another step in advance, giving me as it did the
idea of there being too much chalk, so I went on making different mixtures until I came to 5 of chalk and 2 of Medway clay,
and this gave a result so satisfactory that hundreds of tons of
cement so mixed were soon afterwards made. Some of this
cement was sent to the French Government Works at Cher-
bourg, and was, as I believe, set up as a standard of quality to
which all subsequent purveyors had to conform." ·

**Anticipates Process of Grinding with a Minimum of
Water.**—Mr. Johnson subsequently explains how he partly
anticipated Goreham's patent of grinding with a minimum of
water, and details his experiences as a cement maker on his
own account, first at Frindsbury near Rochester (where he
established the first manufactory on the Medway), then at

* Aspdin, we learn from one of his letters, made his cement from a
mixture of tender and hard-burned clinker. ·

Cliffe, later at Gateshead-on-Tyne, where he followed Aspdin who had failed in business, and lastly at Greenhithe.

Last Traces of the Aspdins.—We think that Mr. Johnson's account is very interesting as the personal narrative of one of the founders of the cement industry, and one to whom, as we shall see later, it owes some important improvements. In bringing this somewhat lengthy chapter to a close we may mention that the elder Aspdin was still working upon a small scale at Wakefield in 1853, when we lose sight of him, while his son William, after his failure at Gateshead,* went to Germany and died in Holstein during the Schleswig-Holstein war in 1864.

* A writer in *The Builder*, in March, 1880, states that, before the success of Aspdin's new material was thoroughly assured, "he launched out by renting several acres of land near Gravesend, and commenced building 'Portland Hall' in such an elaborate and artistic manner as to require some £40,000 to bring it to completion. When the structure was rather more than one-third up, he had to stop, and he sold off at so great a sacrifice that he left the country and died abroad."

CHAPTER IV.

THE COMPOSITION OF PORTLAND CEMENT.

Influence of Materials on the Seat of the Trade.—The inducements originally offered to cement makers to settle in the Thames district are easy to understand, the ample supply of chalk, and the excellent Medway mud, were admirably adapted for their purpose, and, in course of time, a large colony of manufacturers established themselves on the estuary of the Thames, and the London Portland cement obtained a wide reputation. Indeed, the belief gained ground that no really good cement could be made elsewhere, and the raw materials of this district were transported at great expense to other less-favoured localities, in which the manufacture was carried on.

Failures caused by want of Chemical Knowledge.— Owing to the imperfect chemical knowledge of cement makers, the process was at first, as we have seen, exposed to continual dangers, and the failures of the cement made from the lias and from the lower chalk were at one time so serious as to appear likely to bring all other cements into discredit. The constant fluctuations in the composition of the thin beds of stone and shale in the lias formation, and the varying quantities of clay combined with the lower or grey chalk led the cement maker who relied upon rule-of-thumb into all kinds of difficulties and dangers.

Cement can be produced wherever Suitable Materials exist.—It is now perfectly well known that suitable mixtures of carbonate of lime and clay can be prepared from raw materials, to be found in all parts of the world, and that these materials also exist, naturally compounded and ready for immediate use, in vast beds in many places. But until cement manufacturers called the men of science to their aid they too often depended upon some cunning workman, who made a mystery of his proportions, and concealed his ignorance and accounted for his failures by statements which the master took upon trust or did not venture to dispute.

An Early Authority on Cement Making.—We remember such an authority in one of the older works on the Medway, who

had washed whole backs full of over-limed slip (the liquid mixture of chalk and clay), and who attributed the "blowing" of the
resultant cement to an unpropitious state of the atmosphere.
The real fact of the matter being that some new beds of chalk
were being quarried which contained only 5 per cent. of clayey
matters, instead of the 10 per cent. present in the chalk to which
he was accustomed, and he had dealt with the new chalk just
as he did with the old.

Proportions of the Ingredients.—The exact percentage of
clay which, when of the right quality, may be added to the
carbonate of lime to produce Portland cement varies between
somewhat wide and not very clearly defined limits. All mixtures
which contain say from 72 to 77 per cent. of carbonate of lime
will, when sufficiently calcined, produce a Portland cement of fair
quality. When the percentage of clay runs too high—that is,
when the carbonate of lime in the dried "slurry," or mixture of
lime and clay falls below 72 per cent., a cement is obtained which
is technically termed "over-clayed." Compounds of this nature
fuse or "run" in the kiln at a temperature below that required
for the production of a sound "clinker" (the term applied to the
fused material issuing from the kiln). The resulting cement is
light in weight, is apt to set quickly, has a brownish colour, and
never becomes thoroughly indurated. These light descriptions
of Portland, moreover, have a tendency to crumble on exposure
to the weather.

Light Cements well adapted for Plasterer's Work.—
They are essentially plasterer's cements, as they work readily
under the float or trowel, and, owing to the fact that they set
more speedily than a dense cement, the workman can leave them
within reasonable time, which he appears very often to be unable
to do when using a dense slow-setting sample of Portland. These
over-clayed cements are now but rarely made, though at one time
they were much sought after and preferred for certain descriptions
of work. It was the fraudulent substitution of cements of defective quality of this character, prepared from slack-burnt clinker,
for dense Portland which, for a time, had a prominent share in
bringing this excellent material into disrepute. There is a great
temptation to produce these light cements, as they are ground
with comparative ease, and they enable the cement maker to
dispose of his yellow clinker without further calcination.

Over-limed Cements.—The presence of an excess of lime, on
the other hand—that is, where the proportion of carbonate of
lime in the slurry ranges above 77 or 78 per cent. leads to the
production of a cement which will stand the highest temperatures
in the kiln without risk of fusion. Such cements, technically

termed "over-limed," are extremely dense when burnt at a high temperature, they set very slowly, and are very difficult to grind. The sample of Portland produced from a mixture of this kind is apt to be treacherous when used in construction—*i.e.*, unless it has been properly "purged" or exposed in thin layers to the atmosphere, to cause the limey particles to become air-slaked, it is liable to "blow" or swell in the work, an action very frequently caused by the retarded hydration of certain of its particles.

Danger of Heavy Cements.—Looking to the high tensile strength and the great density now demanded under certain specifications, manufacturers are induced to raise to the utmost limit of safety the percentage of lime, and thus to approach more and more to the danger point above indicated, a danger only avoided by extremely fine grinding and careful purging. Apart from considerations of safety, there is a real gain to cement producers in the storage of Portland cement for some considerable period previous to sale. The gain in weight by the absorption of water and carbonic acid from the atmosphere has been found by practical experiment to amount to upwards of 5 per cent.

The Present Position of the Cement Manufacture.— From the point of view simply of materials, we have now entered upon a phase of the Portland cement manufacture, when the trade has passed far beyond the boundaries of the Thames and Medway districts, and when even the lias formation has ceased to present its former attractions to the cement maker. Since, as we have seen, it has become a matter of common knowledge, that chalk is not an essential factor in the production of a good sample of Portland, while, with the growth of skill and chemical knowledge, manufacturers fight shy of the treacherous and uncertain beds of the lias formation, many of the shales in which contain, as already stated, so much iron pyrites and sulphur in other forms as to render them wholly unsuitable for the production of a really first-class sample of Portland. The factories established at various seaports in the north, where chalk could readily be obtained at low freights as a return cargo, have many of them been closed, and the tendency nowadays is rather to seek for a cheap and abundant source of pure carbonate of lime, whether as a hard limestone or a marble, and a good plastic clay, and to establish works in the immediate vicinity of large centres of population, where the cement is likely to be required. This is partly due, however, to the immense progress recently made in the so-called dry-process of manufacture, which, as we shall hereafter attempt to show, has many advantages in its favour.

The Constituents of Cements.—It may be as well, before we attempt to treat of the various processes of manufacture, if we examine somewhat in detail the actual constituents of cement, calling attention, first, to those substances the presence of which in Portland is essential, and subsequently to certain other ingredients which either exert no influence upon the mixture, or which in certain cases, or when they exist in large quantities, may be productive of positive injury to the cement.

Essential Components of Cements.—Careful experiments have proved that the only essential components of a cement are lime, alumina, and silica, which substances we have dealt with in our introductory chapter; but there are, secondly, as we have seen, numerous other ingredients present in Portland, some of them tending to facilitate the cement action, others either inert in themselves or merely taking the place of the essential constituents, and lastly certain substances which we must regard as injurious, except when they are found in very small quantities.

Less Important Constituents of Portland Cement.—The colour of Portland cement is due to the iron, which is always present, and belongs to the second of these categories which includes also magnesia and the alkalies. In the last class we must include the sulphur compounds, as also the carbonic acid and water; these two latter ingredients, if found in any quantity, indicate either that the sample of cement is very stale or that the mixture has been imperfectly calcined. We propose now to discuss briefly the action of sulphur, sulphuric acid, magnesia, iron, and the alkalies.

Sulphur in Portland Cement.—The sulphur of Portland cement is derived mainly from two sources, either from the clay or clay-shales, which are often more or less impregnated with calcium sulphate, and which contain, especially in the case of the lias clays, nodules of pyrites or sulphide of iron, or from the coke employed for the calcination, which, when it is obtained from the gas works, is always rich in sulphur compounds. Gas coke, on account of its relative cheapness, is commonly used by cement makers, and English cements almost invariably contain a certain amount of sulphur derived from this source.

Percentage of Sulphur allowable.—Sulphur, when it exceeds certain not very sharply defined limits, is undoubtedly injurious to Portland cement. Its action upon cements is a matter of very considerable importance, and has scarcely received the attention which it deserves. Some authorities recognising its detrimental effects upon cement have attempted by carefully worded specifications either to exclude it altogether, or to limit the quantity present to 1 per cent. of calcium sulphate, a degree of purity seldom if ever attained in the case of English-made cements. In the German standard rules (see Appendix E, p. 213) the beneficial effect of small quantities of calcium sulphate upon Portland is so far recognised that manufacturers are allowed to employ a porportion not exceeding 2 per cent. in order to confer slow-setting properties.

Selenitic Action.—This action of the calcium sulphate was discovered by the late Major-General Scott, C.B., the inventor of selenitic cement in 1854, and in the so-called Scott's cement, patented by him in 1856, this property was, for the first time, turned to commercial account. We shall treat of this material in a special chapter.

Sulphuric acid, when present in Portland cement, acts, it is believed, in the manner described in the case of Selenitic cement upon the uncombined lime present and deprives it of its avidity for water, causing it to set after the nature of a cement. Very small quantities of sulphuric acid will delay the rapidity of the initial set of Portland cement, and we have found, in the case of a fiery cement, that the addition to it of 1 per cent. of its weight of plaster of Paris would delay the time needed to set from five to seven hours.

Messrs. Dyckerhoff's Experiments.—The annexed table, extracted from the *Proceedings of the Inst. of Civil Engineers*, gives the results of some important experiments by Messrs. Dyckerhoff bearing upon this question. They demonstrate that, together with the delayed setting time, a great gain in strength results from the use of trifling additions of gypsum. In fact, this treatment is equivalent to, and may replace the purging which dense highly calcined cements require previous to use; this statement being clearly proved by the bottom row of figures. General Scott pointed out that the same results might be obtained by exposing the cement for a few minutes to the action of a jet of steam which rapidly hydrates the free lime. Indeed, in one of his numerous patents he specifies this process.

Excess of Calcium Sulphate is Injurious.—Although the action of calcium sulphate in small quantities is thus shown to be beneficial, there can be no doubt but that the presence of this substance in proportions exceeding, say, 4 to 5 per cent. of the volume are injurious, for we must remember that calcium sulphate is comparatively soluble in water, and in damp situations much of it would inevitably be washed out of the mortar, and, moreover, this substance never attains to any considerable degree of hardness.

TABLE SHOWING GAIN IN STRENGTH WHICH RESULTS FROM AN INCREASE IN SETTING TIME.

No.	SAME KIND OF CEMENT.	Time of Setting Neat.	Neat Cement 25 parts water to 100 parts Cement.					One Cement to three of Standard Sand.				
			One Week.	Four Weeks.	Twelve Weeks.	Twenty-five Weeks.	Fifty-two Weeks.	One Week.	Four Weeks.	Twelve Weeks.	Twenty-six Weeks.	Fifty-two Weeks.
		Minutes.										
1.	Original Cement,	20	322·9	405·5	517·9	621·7	701·4	115·2	167·9	237·6	301·6	359·9
2.	The same Cement with ½ per cent. of Gypsum,	210	315·8	456·7	573·4	624·6	651·6	142·3	212·0	338·6	352·8	389·8
3.	The same Cement with 1 per cent. of Gypsum,	600	375·6	507·9	567·7	697·2	781·1	159·3	237·6	311·6	368·5	384·2
4.	The same Cement with 2 per cent. of Gypsum,	840	425·4	543·5	688·6	718·5	806·7	180·7	263·2	304·4	374·2	409·7
5.	The same Cement without Gypsum but kept in store for some months,	630	317·3	449·6	550·6	595·3	618·9	167·9	219·1	318·7	359·9	431·1

Note.—No. 1 weighed about 110·7 lbs. per bushel. No. 5 about 105·9 lbs. per bushel of 40 litres. The breaking weights are given in lbs. per square inch.

Deleterious Action of Calcium Sulphide.—There is, however, another very fatal objection to its presence in the cement, and this is owing to its tendency, when it is found in considerable quantities, to become decomposed in the kiln, leading to the formation of calcium sulphide, a most dangerous ingredient of Portland. The troubles that arise from the calcium sulphide are probably mainly due to its tendency to attack and decompose the iron compounds in the cement.

Iron in Portland Cement.—Iron is found in well-burned Portland cement chiefly in the condition of the lower oxides, but there is nearly always a certain proportion of the peroxide present, and with the iron compounds the sulphur from the calcium sulphide reacts, leading to the formation of iron sulphide. This substance slowly attracts oxygen, and becomes converted into sulphate of the protoxide of iron, which subsequently further oxidises and assumes a brownish-red colour. This action goes on with much greater rapidity in the gauged cement than it does before the cement is made up. Test samples when broken have a greenish tint, much more highly coloured than we should expect in a good specimen of Portland, and the fractured surface on further exposure to the atmosphere turns a rusty brown. If the test briquettes are immersed in water they never harden as they should do, especially if made up with sand, and though at first their colour is greenish, it changes to brown as before on being taken out of the water and exposed to the air. The set of cements of this type is very slow, even when water is very sparingly used in gauging them, and with an excess of water they will scarcely set at all. There can be but little doubt but that the presence of iron in this form hinders the crystallisation of the mass, and cement of this character never attains a high degree of tensile strength, especially under the sand test. Calcium sulphide is also found in cements made from iron slag, and is the chief cause of their peculiar colour and of the slowness of their set. We shall allude again to this matter when treating of slag cements.

Magnesia in Cement.—Nearly all specimens of Portland cement contain a small amount of magnesia, this substance being generally present in the chalk, in the mud from the estuaries of the Thames and Medway, as also in the clay shales of the lias formation. The following analyses of materials used in Portland making by Mr. Charles Spackman, F.C.S., are fairly representative in this respect :—

CEMENT MIXTURES.	No. 1.	No. 2.	No. 3.
Sand,*	2·50	5·57	2·58
Silica, ...	11·83	9·61	11·41
Ferric oxide, ...	1·97	2·42	2·34
Alumina, ...	5·23	3·45	4·80
Iron pyrites, ...	Trace.	...	·43
Carbonate of lime, ...	74·18	75·89	74·09
Carbonate of magnesia, . . .	1·29	1·50	2·61
Sulphate of lime, ...	·18	·16	·21
Potash, ...	·90	·88	·93
Soda, ...	·31	·39	·46
Water, ...	1·82	·61	·43
	100·21	100·48	100·29
* Sand, consisting of—			
Silica, ...	2·27	5·03	2·27
Alumina with trace of ferric oxide,	·23	·54	·31
	2·50	5·57	2·58

These cement mixtures, which all produced fair samples of Portland, come from very different districts. No. 1 is a mixture made at Folkestone from grey chalk and gault clay, No. 2 is a mixture made in the Forest of Dean from limestone and clay, and No. 3 is a mixture made from the Barrow lias quarries. All the samples were dried at 100° Centigrade previous to analysis, but the clay in Nos. 2 and 3 having been dried on hot plates before grinding had doubtless lost much of its combined water.

In the typical clays of the Medway and the Thames estuaries the percentage of magnesia and alkaline matter is often relatively small, though the brackish water may in certain cases raise the proportion of sodium chloride. The following careful analysis from the above laboratory gives the exact composition of Medway mud from Gillingham :—

Silica, ...	38·413	As sand in an extremely
Alumina with trace of iron, .	1·856	fine state of division.
Silica, ...	25·249	
Alumina, ...	14·244	
Ferric oxide, ...	6·744	
Lime, ...	·810	As hydrated silicates.
Magnesia, ...	1·727	
Potash, ...	2·957	
Soda, ...	·773	
Water, ...	3·384	
Iron pyrites, ...	·214	

Silica, ...	·118
Ferric oxide with **trace of**	
Manganese, . . .	·214
Alumina, . . .	·123
Calcium sulphate, . . .	·490
Magnesium ,, . . .	·425
,, chloride, . .	·024
Potassium ,, . .	·524
Sodium ,, . .	2·166
	100·455

Mud dried at 100° Centigrade previous to analysis.

It will be observed that magnesia, potash, and soda are here found in nearly equal proportions.

In the Gault clays used in parts of Kent and Sussex, the composition is somewhat different. The following analysis of the Sussex Gault is given by Mr. Carey :—

Water of combination and organic matter, . . .	3·43
Oxide of iron, ...	4·33
Alumina, ...	13·13
Lime, ...	10·58
Carbonic Acid, alkalies, &c., ...	11·73
Insoluble siliceous matter, ...	56·80
	100·00

We give by way of comparison an analysis of a good sample of river mud used with success for cement making, and of a clay used on the Tyne for the manufacture of Portland; they will serve to show that the composition of the clay may fluctuate between somewhat wide limits. In both cases we have deducted the water expelled at 212° :—

	Wallsend-on-Tyne.	River mud.
Carbonate of Lime, ...	1·90	3·52
Silica (combined), ...	33·39	} 68·09
,, (free), ...	22·44	
Alumina, ...	28·04	16·81
Carbonate of magnesia, ...	2·57	...
Potash and soda, ...	3·94	2·65
Iron peroxide, ...	7·78	8·93
	100·06	100·00

Magnesia Compounds Present.—The magnesia is originally present in combination partly with carbonic and partly with silicic acid, but some clays contain small quantities of magnesium sulphate. In the process of calcination the carbonic acid is expelled, leaving caustic magnesia, a substance which behaves in the cement in a similar way to the uncombined lime. It has, however, much less avidity for water than the lime, and becomes hydrated much more slowly.

Experiments of Mr. Jones on Action of Magnesia.—From some experiments quoted in a paper read by Mr. W. Smith before the Institution of Civil Engineers, to which we shall refer later, and which were carried out by Mr. H. T. Jones, of Aberdeen University, 100 parts of magnesium oxide combined with water at the following rate :—

In twenty-four hours,	. . . .	36·00 per cent.
,, forty-eight hours,	. . . .	38·74 ,,
,, nine days, . .	. . . .	44·30 ,,

The theoretical maximum needed to completely hydrate the above quantity of magnesium oxide was 45·00 per cent. It is not stated at what temperature this oxide was produced, but the hydration was in this case fairly rapid, and the calcination was doubtless effected with far less heat than that needed to burn Portland clinker. Mr. Smith tells us that on adding 5 per cent. by weight of this dehydrated magnesium oxide to Portland cement there was a loss in strength, when tested after nine days, of 20 per cent. Made up with 3 parts of sand, the cement and magnesia broke at 162 lbs. per square inch in twenty-eight days, but no corresponding test is given with unmixed cement. The fact of adding this quantity of magnesia to the sample of cement after calcination does not, we think, throw any light upon the subject.

Time Needed for Hydration.—The time occupied by magnesia in solidifying the water is, as is well known, proportionate to the temperature to which it has been exposed in the kilns. It will readily be seen that this fact indicates a source of possible danger in cements calcined at a very high temperature, for we can conceive of cases in which the hydration of the lime and the combination with silica and alumina to form hydrated silicates may have been completed before the magnesia has taken up its water of hydration. As this action is accompanied by increase of volume, the tardy slaking of the particles of magnesia may lead to the disruption of the cement mortar. This action does actually take place, and in the case of certain cements of French origin, in which a dolomitic lime was used containing as much as 25 per cent. of carbonate of magnesia, the expansion of the mass continued for several months after the mortar was made up and brought the material into so much disrepute that the manufacture had to be abandoned.

Compounds of Magnesia with **Silica and Alumina.**— The compounds formed by magnesia with silica and alumina resemble in every respect the similar silicates and aluminates of lime, and the affinity between the silicic acid and the magnesia is greater than in the case of lime ; but, as M. Fremy has pointed out, there is strong reason to believe that these substances also become hydrated much more slowly than the lime compounds, and from this circumstance it follows that mortars which consist of a mixture of the silicates and aluminates of lime and magnesia may, after the initial set of the lime, undergo molecular movements, due to the hydration of the magnesia compounds, which may further disturb the original cohesion of the mass and lead to disruption.

Magnesia Compounds **are Hydraulic.**—It has, however, been found by laboratory experiments that the silicates and aluminates of magnesia acquire on hydration a superior degree of hardness to the similar compounds of lime, and that they remain wholly unaltered under water.

Influence **of Magnesia on Cements.**—When the percentage of magnesia in the cement is small—say under 3 per cent.—and when the purging or air slaking has been allowed to take place, no injury need be apprehended, but in all dense highly-burned cements, the action of the magnesia must be carefully watched. We do not anticipate any difficulties if the magnesia is combined with sulphuric acid as this compound does not become altered in the kiln.

Roman and Rosendale Cements.—In the tender-burned cements of the Roman cement type and in the Rosendale cements of the United States of America, described at p. 19, the magnesia is found to be quite harmless. It should be remembered while treating of the magnesia that this substance by itself is perfectly hydraulic, and calcined magnesia will set in water and produce a stone-like mass. But for this purpose it should be calcined at a cherry red heat or at a very moderate temperature, and it must be ground to a fine powder before use.

Large amount **of Magnesia present** in American **Cements.**—It will be seen from the annexed analyses of American cements of the Rosendale type that many of these materials contain a large percentage of magnesia in addition to lime and alkalies, and that the proportion of silica and alumina is low as compared with our English quick-setting cements, as, for instance, those made from the septaria of Harwich and the

Isle of Sheppey, some analyses of which are added for the purpose of comparison.

ANALYSES OF QUICK-SETTING CEMENTS.

	Layer No. 9. Rosendale.	Layer No. 13. Rosendale.	Layer No. 17. Rosendale.	Sheppey Island. Septaria.	Harwich. Septaria.
Carbonate of lime, . .	43·30	28·48	40·00	63·00	60·5
Carbonate of magnesia, .	26·04	32·86	39·04	4·2	5·7
Silica, clay, and insoluble silicates, ...	18·52	26·00	11·10	32·00	33·0
Alumina, ...	2·18	4·64	2·52		
Peroxide of iron, . .	1·86	1·86	1·42		
Sulphuric acid, ...	1·96	1·18	0·22	...	...
Potash and soda (chloride),	4·24	4·72	4·06	0·8	0·8
Water and loss, . . .	0·20	0·26	0·26	...	...
Total, . .	98·30	100·00	98·62	100·00	100·00

Recent Controversy Respecting the Action of Sea Water on Cement.—Magnesia as a constituent of Portland cement has lately attained considerable notoriety, in consequence of the partial failure of the great harbour works at Aberdeen, attributed by Mr. Messent, Prof. Brazier, and others, to the action of the magnesia in the sea water. This subject was brought under the notice of the members of the Institution of Civil Engineers in a paper by Mr. Wm. Smith, in 1891, and occasioned a most interesting discussion on cement action. We give in Appendix C, p. 208, a brief outline of the above paper and of the discussion that followed, which traverses the whole ground of the controversy.

Alkalies in Portland Cement.—We have still to glance at the action of the alkalies which we regard as very important constituents of the cement. The only alkalies found in Portland and nearly always present in good samples, to the extent of from 1½ to rather over 2 per cent., are potash and soda. These substances are derived mainly from the clays or clay shales, and while their action is most valuable during the calcination, it is more especially so at the time of gauging the cement, for the alkaline silicates, as it is well known, are the only compounds with silica soluble in water. Their office as carriers of silicic acid to the lime and other sparingly soluble ingredients of the cement has, we think, not hitherto been sufficiently studied, and when we are treating of the set of cements and of the recent theories

4

respecting this much debated question, we shall have to revert to this matter at greater length.

The Investigations of Fuchs.—It will be sufficient here to allude briefly to the investigations of Fuchs, who demonstrated very clearly the influence of these substances. Thus he found that decomposed felspar when treated in the humid way with caustic lime parted with 10 per cent.—that is to say, nearly the whole of its potash ; he found further that substances containing but little alkaline matter, such as pumice-stone and pitch-stone, were deprived of the same by lime, and that both lithia and potash could be separated in this manner from lepidolite (a silicate of alumina, lithia, and potash). By a similar process the alkali is separated from most varieties of clay and transferred to the lime. The solubility of alumina and gelatinous silica in potash renders it more than probable that the alkalies accelerate the hydraulic action of the lime by promoting and facilitating the gradual transference to it of the silica.

Kuhlmann's Experiments.—Kuhlmann carried out numerous experiments with the alkaline silicates constituting "water-glass," which established the fact that these materials part with their silica to the lime without the aid of heat, and the same reaction occurs with carbonate and sulphate of lime as with caustic lime —the alkali transfers its silica to the lime and unites with the carbonic or sulphuric acids. Chalk powder exposed to a solution of silicate of potash becomes after a short time solidified into a stone-like mass, and lumps of chalk immersed in the liquid receive a surface coating of silicate of lime, which on subsequent exposure to the air becomes so hard that it is capable of taking a polish. The induration of cement concretes by means of the so-called "silica-bath" depends upon this property, and the preservation of wall surfaces by a coating of water-glass is another illustration of this principle.

CHAPTER V.

PROCESSES OF MANUFACTURE.

HAVING thus glanced at some of the chief components of Portland cement, we may pass on to the more detailed account of its manufacture. The plan formerly adopted was, as we have seen, to incorporate together the chalk and clay, and, at the same time, to bring them to a fine state of subdivision in the presence of water in the contrivance known as the wash-mill. This is termed the wet process, and the piece of apparatus in question was borrowed from the brickmaker, who, long before Portland was produced, had been in the habit of tempering his clay by a piece of mechanism of this kind.

The Wash-mill.—The old-fashioned wash-mill consists in its essential features of an annular trench or pit lined with stone or brick; on the central mound is a pier carrying a vertical shaft with a revolving arm, to which are attached knives, cutters, or, in some cases, harrows, reaching nearly to the bottom of the pit, and serving to break up or disintegrate the clay or chalk. These substances are fed into the pit either continuously or from time to time at stated intervals, together with such a supply of water as to keep the mill constantly at the overflow level. By means of gearing, the shaft is caused to revolve from twelve to twenty times per minute, and the contents of the pit are thus violently agitated; the clay is broken up and the lumps of chalk are comminuted by attrition and ground to pieces. These substances are also evenly distributed throughout the water flowing into the pit, and they are converted into a creamy liquid called, as we have already stated, the "slip," which is led away in wooden troughs to the reservoirs, technically termed "backs."

The Backs.—These backs are frequently of large size, being sometimes as much as 100 feet in length by 40 feet in width, with a depth of about 4 feet. A large back would thus contain 160,000 gallons of the slip. In order to equalise the work it was usual at one time to drag heavy stirrers backwards and forwards along these reservoirs, and thus to "lute" or mix one day's washed clay with the mixture made the day before.

The Semi-wet Process of Goreham.—The large volume

of water which was involved in this so-called wet process had, of course, to be got rid of before the materials could be calcined, and the use of this water entailed, moreover, sundry disadvantages and difficulties, which have been only partially obviated by the modern system of grinding the material with a very much smaller quantity of water. This latter plan, to distinguish it from the first adopted, has been called the "semi-wet" process, and there still remains a third alternative, where the amount of water is reduced to a minimum, and this is known as the "dry" process. It will scarcely be necessary to deal with each of these systems of manufacture in detail, because they all aim at the same result, the thorough incorporation of the carbonate of lime with the requisite proportion of clay. The wet process is in all the more modern factories gradually giving way to the semi-wet process, which, because it was patented some years ago by Mr. W. Goreham, is known under his name. The reduction in the quantity of water has no doubt many advantages in its favour, not the least of which is the fact that none of the soluble constituents are removed, as may be the case when the superabundant water is run off from the top of the backs by means of the peg-board. Then the risk of unequal settlement is, to a great extent, avoided, and there is an undoubted saving of time.

The **Semi-dry Process.**—When we have to deal with limestones already rich in clayey matters, like those of the lias formation, there can be no hesitation in adopting a so-called semi-dry process; and when this process is intelligently carried out a cement is produced of very high quality and at a relatively small cost. In this case the water is only used in sufficient quantities to moisten the materials and to enable them to be pugged and subsequently moulded into bricks or blocks. It will be obvious that for use in this way the mixture must be brought into a plastic condition and when well tempered it may conveniently be converted into wire-cut bricks by means of suitable machinery.

Double Kilning.—There is still another form of dry process; under which the materials may be absolutely dry, and in which they are compacted by means of a heavy blow in a screw or lever stamping-press. Such a process as this might be used when a hard limestone is first calcined, then slaked into powder, intimately mixed with well-ground clay, and finally stamped into blocks for burning. A plan of this kind in which the lime passes twice through the kiln is technically known as "double-kilning." It is of course a relatively costly process, and one which could only be carried out in localities where fuel is extremely cheap; it has the advantage of furnishing the lime in an exceedingly fine state of subdivision, infinitely finer in fact than could possibly be reached by any merely mechanical process. No finer powder is known in nature than that of fully slaked lime, and such a powder if really intimately mixed with levigated clay would give us a theoretically perfect cement "slurry" (this being the name technically employed for the dried mixture of lime and clay). Stamping-presses, such as are used in the potteries for the manufacture of "dust-tiles," would compact this material into blocks or slabs, and some authorities have expressed an opinion that Portland cement of fine quality might be manufactured in this way. It must, however, be remembered that the slaked lime is a much less dense material than the carbonate, and that the heat needed to bring about the combination of the lime and silica is vastly greater than that required for the dehydration of the slaked lime. We have been told by those who have carried out some experiments in cement making on this plan that it proved an utter failure, owing to the fact that the blocks fell to powder in consequence of the expulsion of the water at a temperature far below that capable of effecting any chemical reaction between the lime and clay, and the powder thus produced choked the draught and clogged up the kiln. We shall in this work consider the term "dry process" to apply to those systems of manufacture in which a minimum of water is used, merely sufficient, in fact, to damp the materials and to enable them to be pressed into blocks or lumps by means of powerful machinery.

Cement from Lias Limestone.—The process of making Portland cement from a mixture of finely ground limestone and clay shale in this latter way has long ago passed beyond the experimental stage, and much excellent cement, both in this country and abroad, is made by this means. For instance, Portland cement made from the lias limestone and the clay shales of the same formation has been most successfully produced by a process in which the amount of water employed is reduced to the lowest possible limit. In carrying out this manufacture great care is manifested in selecting beds of stone which, while they contain a fairly large percentage of clayey matters, are comparatively free from sulphuric acid or sulphides.

Great Variations in the Lias Materials.—The percentage of silicate of alumina mixed with the carbonate of lime fluctuates in the lias formation in a very puzzling way; two beds of stone within a few inches of one another will often vary as much as 10 per cent. in their composition. The beds of shale found between the different "floors" of stone frequently exhibit equally wide fluctuations in their contents in lime, and for this reason it is absolutely essential to obtain accurate analyses of each layer in the quarry before deciding upon the proportions of the stone and shale to be employed. Many of the lias shales are moreover so rich in sulphur compounds as to be practically quite unfitted for cement making, and it becomes necessary to seek elsewhere for purer beds of clay. The following partial analysis by Mr. Spackman of the different strata in some lias quarries will serve to illustrate the foregoing observations.

DETERMINATIONS OF CARBONATE OF LIME PRESENT IN A VERTICAL
SECTION OF A LIAS QUARRY.

Floor.	Percentage of CaCO₃.		Floor.	Percentage of CaCO₃.
No. 1,	Stone, 75·20		No. 5,	Shale, 51·71
,, 2,	Stone, 83·12		,, 6,	Stone, 88·22
,, 2,	Shale, 56·87		,, 6,	Scull, 68·54
,, 3,	Stone, 85·00		,, 6,	Shale, 49·02
,, 3,	Scull, 78·39		,, 7,	Stone, 84·49
,, 3,	Shale, 39·86		,, 7,	Scull, 69·59
,, 4,	Stone, 86·15		,, 7,	Shale, 51·34
,, 4,	Shale, 48·59		,, 8,	Stone, 81·83
,, 4,	Shale, 50·60		,, 8,	Scull, 77·13
,, 5,	Stone, 88·14		,, 8,	Shale, 48·42
,, 5,	Scull, 69·65		,, 8,	Shale, 58·12

Note.—The shales alternate with the layers or floors of limestone and are of varying degrees of hardness; those containing the highest percentage of carbonate of lime being the hardest. Adhering to the stone, but separating easily from the shales, are layers of a substance about 1 inch in thickness, locally known as "scull." This is carefully chipped off the stone used for the preparation of lump lime and is burnt separately for making ground lime.

Cement Manufacture.—Having selected the raw materials and dried the shale or clay on hot floors, to expel the superfluous moisture, the subsequent process of manufacture by the dry process is as follows :—The materials are crushed between rollers, or by means of a stone-breaker, and ground under millstones to a very fine powder. This grinding is of great importance as it must be relied upon, not only to bring the particles of stone and clay into very intimate admixture, which experience has proved can best be done in this way, but also to reduce the particles of limestone to such a degree of fineness that they may readily approach the silicates in the clay when the requisite temperature is reached in the kiln.

Dry Materials Need to be Slightly Moistened.—For convenience of handling, and for the purpose of drying the mixture and filling the kiln, it is usual to compress the ground materials into blocks, and in order to do this the powder from the stones will require to be slightly moistened and then passed into any of the so-called dry-press brick machines to be moulded. It is not necessary here to illustrate the various types of machines which have been employed for this purpose. Some manufacturers prefer one machine and some another, but as the manufacture of bricks by this means is a recognised branch of industry, and has attained large dimensions in many parts of the country, the intending cement maker will do well to select that type of machine which is best understood in the locality in which he is working. There are many little niceties to be observed in the damping of the clay, the feeding of the machine, and the degree of pressure to be employed, which cannot well be explained, but which really come of experience, and must be adjusted to the character of the raw materials. In the arrangement of the works it will be important that the various processes should be carried on consecutively, and that the amount of handling should be as far as possible reduced to a minimum.

Importance of Labour Saving in Cement Making.—It is, therefore, necessary to carefully study the different levels at which the machines are placed, so that the dust from the stones may be delivered at such a height that when damped or moistened it can be at once transferred into the brick-making machine, and the bricks should, if possible, be produced at such a level that they may be wheeled direct from the machine on to the drying floor or into the chamber, without any more considerable gradients than the workmen can easily surmount. It is in the skill with which all these small matters are arranged that the possibility of making a profit arises, especially in these days when there is such a severe competition in all branches of the trade.

Utilisation of Waste Heat.—It is, moreover, important to utilise to the utmost the waste heat from the kilns in drying the cement materials, and in many of the older works very little attention was devoted to these matters. Now the subject has been attacked by a host of inventors, and we have quite a number of patents which aim at turning the kiln-gases to advantage.

Johnson Chambers.—Chief among the patented processes of using the waste gases from the kilns for the purpose of drying a fresh charge of slurry is that of Mr. I. C. Johnson, one of the pioneers of the cement manufacture, whose reminiscences we have quoted somewhat extensively in Chapter III. We shall describe his invention when we come to treat of drying processes in a future chapter. Following somewhat on the lines laid down by Mr. Johnson are the modifications suggested by Mr. Michele, Mr. Gibbons, and others, who either extend the area of the drying surface by using the arch of the chamber for this purpose, or who pass the kiln gases under as well as over the layer of slurry which has to be dried. We shall dwell on these matters more in detail when discussing the different forms of drying chambers.

Process of Cement Making Advocated.—It will perhaps be expected that we should here point out what we consider to be the best process for the manufacture of cement, but the answer to this must to some extent depend upon the nature of the materials to be employed. If we have to deal with soft materials like chalk and clay, the semi-wet process has much in its favour, as it is certainly the most economical system of dealing with substances of this kind.

Process Best Adapted for Hard Limestones.—The foregoing process is, of course, as far as the wash-mill stage is concerned, out of the question when we have to employ hard limestones and indurated shales. The only really satisfactory way of dealing with materials of this character is by means of a dry-process, for we do not need any large volume of water to enable the ground limestone and the clay to be incorporated together into bricks or blocks or moulded into lumps, in order that they may be stacked in the kiln. It will, therefore, be evident that in the case of chalk and clay we have the choice either of a semi-wet process, or, if we avoid the use of excess of water, of a semi-dry process, while, in the case of hard limestones and shales, we are confined to the latter system of manipulation. Cement of the most excellent quality can be made by either process, but of the two, in point of speed, compactness of works, and economy in the manufacture, we prefer the latter system. There can be no possible doubt that some of the best cements now in the market are made by this process, and this plan of producing Portland cement is almost universal on the Continent.

Relative Economy of Dry Process.—A careful comparison of the cost of labour and fuel for the two methods now under consideration enables us, moreover, to pronounce for the dry process in point of economy, and this is the aspect from which the question must most undoubtedly be approached by the manufacturer in the future, as the selling price of Portland cement has latterly been so much reduced that profits are cut down to a very low figure. We have dealt in Appendix F, p. 218, with the cost of cement-making in various parts of the country.

Situation of Works.—As a matter of fact, the question of the profitable production of Portland is fast becoming one of facility of railway communication and cheapness of carriage. The manufacturer must not only consider the advantages to be secured by a cheap supply of raw materials and of fuel, but he must take into account the proximity to the markets, and the all-important subject of railway rates. In the London district the cement manufacturers of the Thames and Medway must undoubtedly be able in any case to hold their own, but the competition with makers in the North of England is likely to be a fierce one, for the trade in the North and in the Midlands, especially when the advantages of the dry system are more fully realised.

Importance of Uniform Test.—We look forward with confidence also to the time when the adoption of some uniform system of testing will have induced our manufacturers to pay as much attention to fine grinding as is now the case on the Continent, and when it will become possible to employ Portland cement with large proportions of sand for many of the purposes for which ordinary lime is now used. This will no doubt give an immense impetus to the trade, and will enable cement manufacturers to furnish a greatly improved material at a remunerative price.

CHAPTER VI.

THE WASH-MILL AND THE BACKS.

Manufacturing Processes based on Brick Making.—It seems highly probable that the cement maker in the early days of the manufacture availed himself, as we have seen, of many of the processes of the brickmaker, who, long before Portland cement was thought of, had learned to prepare his clays, to add the necessary amount of chalk, and to settle the slip in reservoirs much in the same way that the cement maker sets to work, adopting, in fact, processes which date back to the period of the Pharaohs. At any rate, in the London District which, as already stated, soon became the chief seat of the Portland cement trade, the brickmaker had for generations made use of the wash-mill and the reservoir or back for the slip, and these contrivances remained until comparatively recently in common use at cement factories, but little changed or modified. There can be no doubt that for soft materials like the grey chalk, and for the alluvial river mud, the wash-mill is an excellent contrivance, but when we have to deal with compact shales like some of those of the lias formation, or with the hard, flint-bearing upper chalk, even the best arranged wash-mill leaves much to be desired.

Original Process of Portland Cement Making.—In the earliest Portland specification, that of Aspdin, to which we have already referred, the washing together of the ingredients was not required, as he used a hard limestone, and, in order to reduce it to powder, he evidently calcined it and slaked it, for we cannot imagine that he could have relied upon obtaining any large supply of his materials from the road scrapings of the vicinity of Wakefield, or that he ground the raw stone. He says, however, "I take it from the roads, after it is reduced to a puddle or powder," and he then goes on to explain that, failing this puddle, he takes "the limestone itself." It is somewhat obscure why he calcines the puddle or powder, and perhaps in this respect his specification is wrongly drawn. At any rate, with his lime in a state of powder, he incorporates "a specific quantity of argilla-

ceous earth or clay"; these materials he mixed "with water to
a state approaching impalpability, either by manual labour or
machinery." Whether he used a wash-mill or a simple "blunging-
mill," such as is employed by potters to bring their clay into
"slip," is not clear, but we incline to the probability that it was
the latter form of apparatus, for he goes on to use the term "slip-
pan," which is certainly a piece of pottery plant, and there were
potteries in operation at that time in the neighbourhood. In the
slip-pan, by means of natural or artificial heat, he dries the mixture of lime and clay "till the water is entirely evaporated, and
it is subsequently calcined in a furnace similar to a lime kiln.
Here we have an outline of a process, which has, as we have
pointed out, been called "double kilning," for the limestone is
burnt twice over—first, to enable it to be slaked to a fine powder,
and secondly, after it has been incorporated with the clay, to
burn the mass into a clinker.

Fine Grinding of Raw Materials.—It is interesting to
notice here that Aspdin insists upon the reduction of the
ingredients to a "state approaching impalpability," and there
can be no two opinions respecting the importance of this pre-
caution. Some of the recent improvements in the wash-mill
aim at a more carefully regulated reduction of the materials than
is possible in the mill in its crude form.

The Common Wash-Mill.—We have already described the
early wash-mill, and may, therefore, now pass on to the more
recent form of this apparatus. It will have been evident that
the mill first used was a most clumsy contrivance, as the feeding
is at best intermittent, moreover, all the heavy impurities—flint,
pebble, &c.—sank to the bottom and tended to clog the action
of the cutters.

Wastefulness of Wash-Mill.—The large percentage of waste
in the wash-mill when hard chalk was used led manufacturers in
certain cases to employ edge-runners as an adjunct to correct
the imperfect attrition. The slip passing from the mill was
conducted into the pan of the edge-runner and here all the hard
particles of chalk were crushed. In this way it became possible
to utilise the waste deposited in the shoots and intercepted in
the catch-pits, and a much more uniform sample of slip was
turned out. The accumulations at the bottom of the pit involved
frequent stoppages and emptyings of the wash-mill, especially
when a flinty chalk was being used. As the charges of chalk
and clay are not always fed in at equal intervals of time, even
if the weights or measurements of the ingredients are fairly
uniformly adjusted, it follows that the slip is never quite the

same in composition, though this is all made right in the great back or reservoir into which the liquid is conducted.

Imperfect Mixtures Corrected in the Backs.—It used formerly to be the aim of the manufacturer to construct backs of increasingly larger capacity, so that any trifling fluctuations from day to day in the composition of the slip might be equalised and corrected by "luting" the back. The specific gravity of the particles of chalk and clay differ in some cases rather widely, and this would be apt to lead to unequal settlement or a tendency to deposit in layers.

Methods of Intercepting Coarse Particles.—Where the wash-mill is urged at too great a speed, fragments of hard chalk which have escaped perfect attrition are frequently washed out into the troughs, and all sorts of devices are made use of to intercept them. The slip is made to flow through gauzes or fine gratings of wire, and numerous catch-pits or depressions are arranged in the channels. In some cases the "shoots," or channels, are made to traverse a long and devious course, so as to give every possible chance to get rid of these troublesome particles.

It will be obvious from the foregoing description of this stage of the cement manufacture that it is of great importance to arrange the levels of the wash-mill, backs, &c., so that all the operations can be conducted by gravitation. The wash-mill should be placed at a high level and should be handy for the charges of the raw materials, the chalk from the quarry and the clay from the barge. In many of the old-fashioned works too little attention seems to have been paid to these details. We often find that the slip has to be lifted from the wash-mill, because it is usual to place this machinery close to the wharf, whereas it would entail far less labour if it had been placed on high ground at the back of the works, handy for the chalk, and involving only the wheeling of the clay. These matters seem, however, to have been but little studied.

Michele Wash-Mill.—Mr. V. D. de Michele patented in 1877 an improvement in the wash-mill which gave very good results. In his invention the liquid slip, instead of flowing out at one spot in the circumference of the washing-way, is made to pass over a flat lip which entirely surrounds the mill. The arrangement of the Michele wash-mill will best be understood by reference to the accompanying illustration. (See Fig. 2.)

It will be seen that in its essential features this mill combines the action of the wash-mill and that of the edge-runner, for while the materials are in the first instance reduced by attrition to a liquid slip in the pan of the wash-mill, the slip in escaping

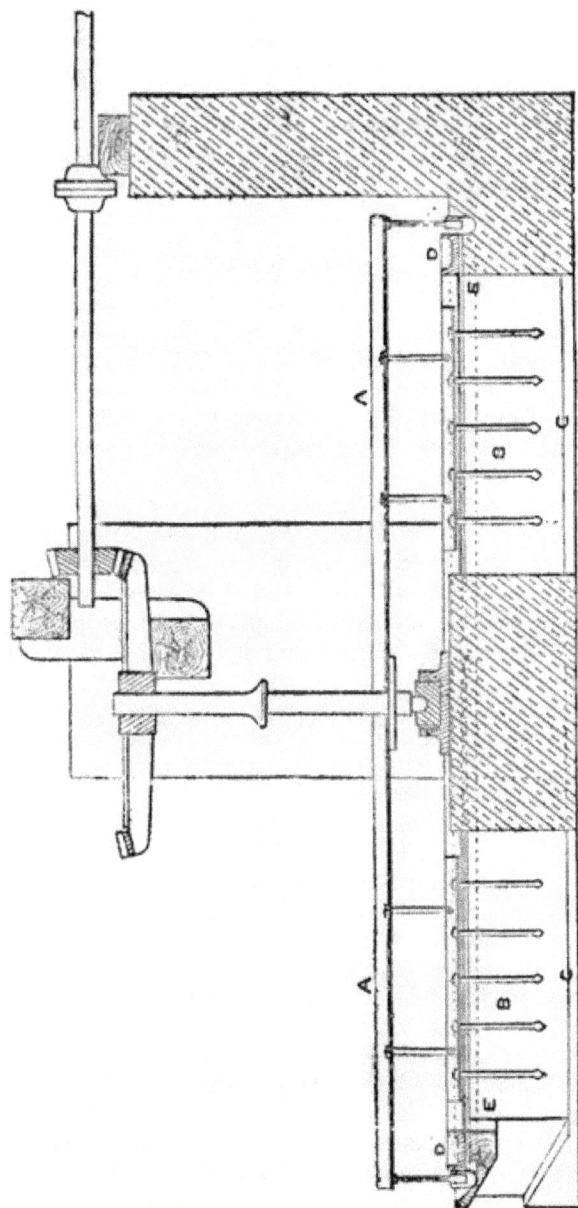

Fig. 2.—The Michele Wash-Mill.

over the edge is ground between the metallic surfaces, and all
hard particles are broken up. The grinding surfaces of metal
are carefully adjusted, and friction is not excessive, because the
runner, or upper plate, is counterbalanced and is acting upon
small particles of chalk and clay which serve as lubricants.

The Goreham Process.—All these systems of washing the
ingredients into a very liquid slip have, however, recently been
to a great extent superseded by the process of wet-grinding with
a minimum of water, which was patented by Mr. W. Goreham
in 1858, but which has been used in France for more than 100
years at Meudon, in the neighbourhood of Paris, in the preparation of hydraulic lime from a mixture of chalk and clay. The
original inventor of this milling process is said by Mr. Reid to
have been M. Dupont. Mr. Goreham specifies the use, in the
first instance, of the wash-mill, to incorporate together the chalk
and clay, but with only about one-fourth of the water at present
required; and having made a mixture in this way, it is passed
between horizontal mill-stones and ground in a creamy state,
having just so much water with it as will permit it to flow
gradually through the shoots on to the drying floors or plates.
In France the edge-runner mill is preferred for this purpose, and
in some English cement factories edge-runners are employed in
lieu of the burr-stones, but the grinding by this means is said to
be less uniform.

Advantages of Goreham's Process.—In large works this
process leads to a great economy of time and much saving of
space, as the backs are entirely dispensed with. Another advantage is that the consistency of the mixture is such as to
reduce to a minimum the chances of irregular deposition of the
particles of chalk or clay, due to the difference in their specific
gravities. The Goreham process is well adapted for use in
conjunction with the arched drying floors attached to the kilns
patented by Mr. Johnson, and the use of what has been termed
the "semi-wet" process is being constantly extended.

The use of Backs.—We have said but little concerning the
backs which are used for the storage of the slip, they are generally
placed at the rear of the works convenient of access to the drying
floor, on to which the consolidated slurry is wheeled in barrows and
spread in layers of uniform thickness over the plates. The sides of
the backs are generally constructed of brick, rubble, or concrete
walls about 4 feet in height, and the bottom is carefully formed
of porous materials, dry ashes, or gravel to permit of the ready
soakage of the water. Special attention should be paid to the
subsoil drainage. The suspended chalk and clay, when they

come to rest in the back, rapidly subside ; part of the water soaks
into the ground or evaporates, but a considerable quantity of
the clear supernatant water is drawn off by means of the " peg-
board," which is a perforated penstock, having numerous small
openings closed by pegs ; the holes being at such levels that the
water can be run off gradually as far as the top of the slurry.
Some weeks must elapse, even in the summer time, before the
contents of the back become sufficiently consolidated to dig them
out, and to wheel them in barrows on to the drying floors.

CHAPTER VII.

FLUE AND CHAMBER DRYING PROCESSES.

Drying Floors.—We have seen that in the semi-solid form the
cement mixture is known as "slurry," and it will then contain
from 40 to 50 per cent. of water. The drying floors for expelling
this water are of many different kinds, some makers are in favour
of flues covered with cast-iron plates, some substitute slabs of
fireclay, or they use fire-lumps at the hot end above the furnaces
and plates beyond. The heat is supplied either from slow combustion furnaces, employed solely for this purpose or by coke-
ovens, or the waste heat from boiler-furnaces and from the
kilns may be utilised. As already stated many different plans
have been proposed in recent times for taking advantage of the
waste heat from the other parts of the works for the drying of
the slurry.

   The Drying of Slurry.—The desiccation of the slurry
requires care and attention, as it is advisable not to overheat the
flues, and thus to cause the temperature to rise to the boiling
point. If this action is accidentally permitted to take place the
slurry is rendered very porous and friable, and likely to crumble
when interstratified with fuel in the kilns. The more slowly and
steadily the drying is effected the denser and better will be the
resultant material. In some of the more modern systems of
drying this substance the hot air is caused to pass over as well
as under it, and in the plan patented by Mr. I. Johnson the
surplus kiln heat passes over the slurry in specially arched
chambers attached to the kiln.

   Patent Systems of Drying.—Following on the lines of
Mr. Johnson's patent, a number of inventors have proposed
in various ways to utilise the waste heat from the kilns for the
drying of the raw materials, and we shall attempt to indicate
briefly some of the various systems that have been introduced
for this purpose. It will readily be understood that with the
keen competition which now exists in all branches of this industry it is most important to save as much as possible, the
heavy expense entailed in the drying of the slurry and in burning
the clinker, the cost of which has been sometimes set down as

high as 7s. per ton of the finished cement. It is not an easy matter to separate the cost of the drying and calcination, for the estimates given by manufacturers vary between some wide limits, but there is no room for doubt that in old-fashioned works these processes are carried out in a very wasteful manner.

Coke-Ovens.—Some of the chief modifications which have been brought about recently in cement manufacture relate to this stage of the process. The first attempts made to save fuel were those based upon using the waste gases of the coking process, and coke was produced by the cement manufacturer on an extensive scale and at a relatively high cost for the sake of this waste heat. It is difficult to obtain any reliable estimate of the extra cost entailed in making coke in this way, as compared with the purchase of the relatively cheap coke, which may be regarded as a bye-product in gas manufacture.

Gas Coke.—In many parts of the country gas coke is almost a drug upon the market, and though it is not quite the best of fuel for the cement maker, he is able to use it largely, especially if his raw materials contain but little sulphur. There are cases, however, in which it is scarcely expedient to employ a fuel so rich in sulphur compounds as are many descriptions of gas coke, and an objectionable amount of ash sometimes results from the impure coke. In some figures kindly communicated to the author by Mr. Johnson, he gave an estimate of the relative cost of gas coke and coke specially produced in cement works, as follows :—

Coke-oven coke, . . .	16s. 10d.	per chaldron.
Gas coke, ...	10s. 6d.	,,

and this is probably a fair and reasonable proportion. It is well known that coke specially made for engine and foundry purposes is a much more costly fuel than that prepared in gas works, and an extra expense of 50 per cent. is quite what we should expect for coke made in cement works.

Mr. Johnson's figures relate to experiments carried out to demonstrate the advantage obtained by the employment of drying chambers for the slip attached to the kiln, to which process we have already referred.

Coke made in Cement Works.—Coke produced in suitable flat ovens for the cement maker's purpose is never so good in quality as that prepared in ordinary vertical ovens, such as are used in the colliery districts for the manufacture of engine coke. The reason of this is not far to seek, the combustion of the coal takes place in too confined a space, and the coke rarely obtains an adequate supply of air to burn off the impurities. The object

5

of the cement maker is in fact not to turn out a first-class sample of coke, but to heat his drying floors. Again, the coal employed is not always a rich bituminous coal suitable for coking purposes, but care should be taken to avoid coal yielding a high percentage of ash, as this may prejudice the quality of the cement.

Waste in Coke Making.—It seems almost impossible to prepare a lustrous silky coke in any other ovens than those in which the products of combustion pass freely away into the atmosphere as waste products. All attempts so far to utilise the gases evolved in the coking process for heating purposes have inevitably resulted in reducing the quality of the coke.

Position of Coke-Ovens.—For the purposes of the cement maker the coke-ovens must be conveniently placed for loading the kilns, so as to avoid as far as possible expensive handling. They should not be of too large a capacity, as the charges are generally withdrawn once in twenty-four hours. Ovens holding from 20 to 30 cwts. of coal are preferred. Such an oven would be about 9 feet long, 4 feet 6 inches wide, and 3 feet 6 inches to the crown of the arch. The coal will waste as much as 50 per cent. in weight in the coking process, and it is important to so arrange the flues and to control the draught by means of dampers as to utilise as much as possible the products of combustion in heating the flues. The ovens are necessarily constructed of fire-brick, and small apertures are left in the arched roof for the escape of the gases which burst into flame on mixing with the air, and pass along the shallow flues of the drying floor to the chimney at the further end.

Flue Construction.—Practical men vary very much in their estimate of the right length to make the flues, which to some extent must depend upon the draught and the height of the chimney, though from 60 to 70 feet seems about the usual length from the front of the ovens to the collecting flue at the end. The floor above the ovens, and for about 12 feet beyond them, should be formed of fire-clay materials. Special tiles are made for this purpose, and the remainder of the floor generally consists of cast-iron plates about half an inch in thickness. It is bad economy to make the flues too shallow, as they speedily become choked with soot and dust. The longitudinal flues, 12 inches by 9 inches, with midfeather walls of 4½-inch brickwork, pass into a cross flue at the end leading into the chimney. Proper dampers should be provided to regulate the draught which may conveniently be placed where the cross flue enters the chimney. It is very difficult to regulate the heating of the flues when connected with coke-ovens, and there is a great wear and tear, due to

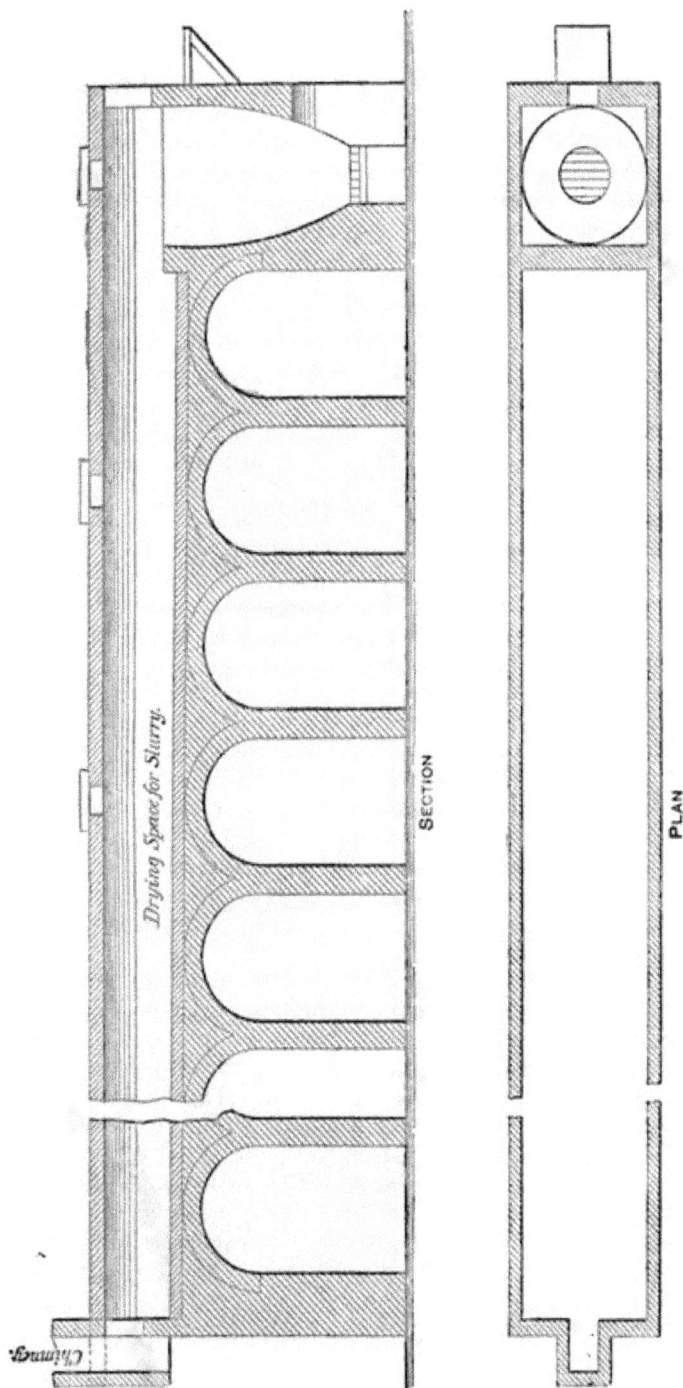

Drying Space for Slurry.

SECTION

PLAN

Fig. 3.—Johnson's Drying Chambers.

altered expansion and contraction, of the plates and flue-covers used in this way.

Coke Making in Cement Works is Declining.—The practice of making coke on the works for the sake of drying the slip by means of the heat evolved during the process has, we believe, become almost obsolete, and in works of modern construction this very costly plan is no longer thought of. Small coal of the cheapest description can be used to greater advantage for the flues if this system of drying is resorted to, but even drying on flues has been abandoned for the drying chambers attached to the kilns on the Johnson or some similar system. By this means the great wear and tear of the plates is avoided, and much expensive handling of the raw material is saved. Moreover, it is claimed that under this system, where the heat passes above the slip, a more tough and dense slurry is obtained than from the hot floors with flues beneath them. The explanation given for this is, that the heat is often sufficient, as we have seen, to cause the slip to boil, and when such is the case the material becomes filled with minute spaces caused by the steam bubbles. There can be no doubt that the slurry produced in the chambers on the Johnson system is a much tougher and more dense substance than much of that dried in the old-fashioned way on flues. The density or toughness of the slurry is a matter of considerable importance to the cement maker, especially when large and lofty kilns are employed. If the slurry in the lower part of the kiln crumbles and becomes much crushed, it is liable to interfere with the draught and entails the production of an undue proportion of faulty and underburnt clinker.

The Johnson Chambers.—The plan patented by Mr. Johnson consists merely in the addition to the kiln, at or about the top level of the cup-shaped portion, of an arched chamber of the same width as the kiln, and of such a length that the floor area will hold enough of the wet slip to serve when dry as the charge of the next kiln. As will be seen from our illustration (Fig. 3) this chamber is simply a short tunnel, interposed between the kiln and the chimney, and each chamber may be from 12 to 14 feet in width by from 80 to 100 feet in length. The dimensions will depend upon the size of the kiln and the height of the chimney. The floor is slightly inclined, falling towards the kiln, so that the stratum of slurry may be deepest where the greatest heat is found in the vicinity of the kiln; the depth of the layer of slurry may vary from 10 to 12 inches, close to the kiln, to about 3 or 4 inches at the far end of the chamber. The use of these chambers involves a considerable extra chimney power, and lofty chimneys must be erected in order to produce the best results.

Sulphur Deposits on Surface of Slurry.—An objection which has been raised against this plan is, that with some descriptions of slurry there is a very considerable deposit of solid particles, chiefly sulphur compounds, upon the surface of the matters in the chambers. The hot gases passing out of the kiln are condensed when they come in contact with the cold slurry, and we have seen in some cases the whole surface heavily coated with such a deposit. The best answer to this objection is that no permanent injury is caused to the resultant cement, as nearly all this matter will again sublime when it comes into the kiln, and is exposed to the full heat during the process of burning.

Gibbons' Patent Chambers.—In a plan of drying patented by Mr. R. A. Gibbons, which is shown in the accompanying illustration (Fig. 4), the hot gases from the kiln pass both under and over the substances to be dried, and by this means it is claimed that a larger amount of heating surface is obtained, and the sublimed sulphur compounds are less liable to be deposited on the top of the slurry. Two or more chambers are employed which are used alternately, the heat being passed both under and over the material to be dried.

Michele's Chambers.—In a somewhat similar way Mr. V. de Michele uses both the exterior of the arches as well as the floor of the chambers for drying the slurry. For this purpose the arches are very lightly constructed, and, as will be seen from our illustration (Fig. 5), which gives a section through a series of chambers, they have a roof above them to protect them from the weather, so that an additional quantity of the raw material may be dried on the top of the chambers. It is necessary, however, to spread the slurry on these arched surfaces by hand, and this must, we think, entail a considerable amount of labour.

Spackman's Drying Chambers.—When the cement compound is made into bricks or blocks, the drying chambers attached to the kilns on the plan invented by Mr. C. Spackman and shown in the illustration (Fig. 6), may be used with considerable advantage. It will be seen that, in this invention, the chambers are rectangular in form, and are placed alongside the kilns at such a level as to be most convenient for the loading operations. The kilns, which are of the usual type with a conical chimney, are burnt in the ordinary way with inter-stratified fuel. The orifice of the cone is provided with a damper, and when the charge is in full fire and the combustion is nearly completed the upper damper is closed, and the dampers in connection with the underground flues leading to a lofty chimney are raised. By this means the hot gases from the kiln

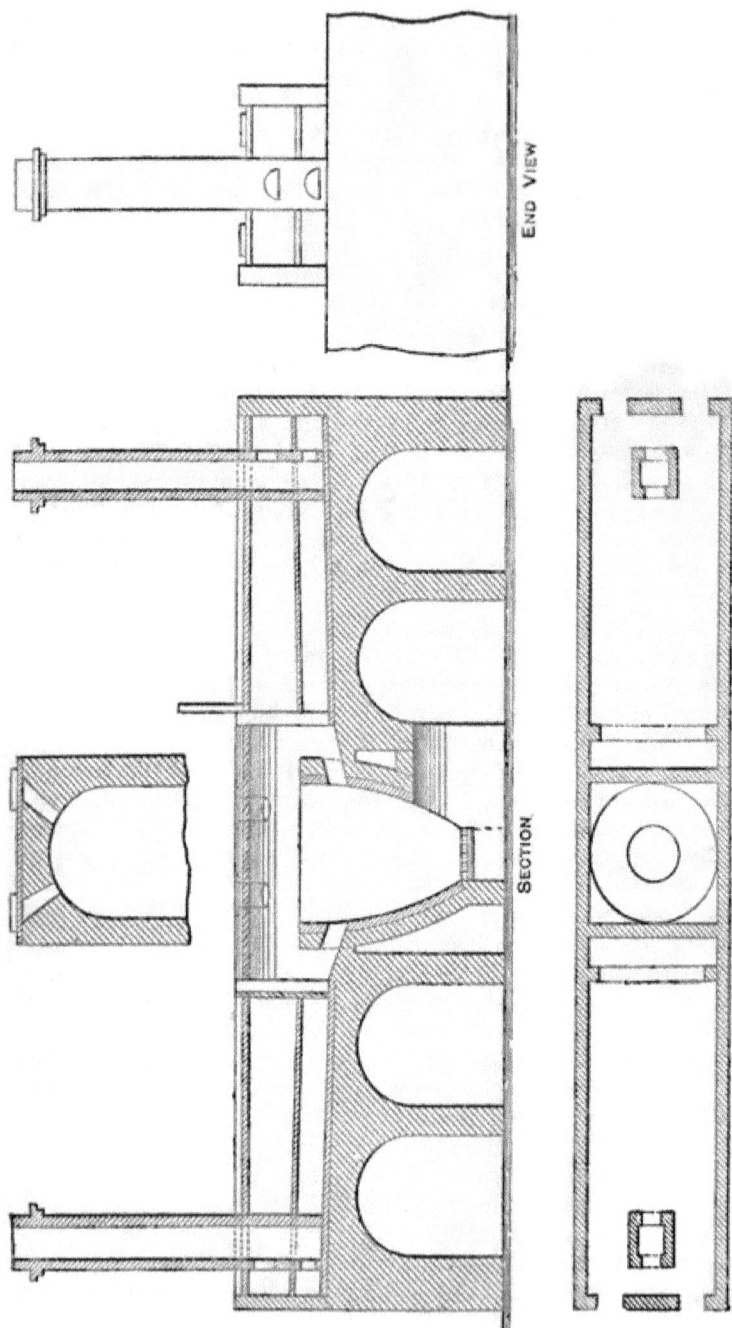

END VIEW

SECTION.

PLAN

Fig. 4.—Gibbons' Patent Chambers.

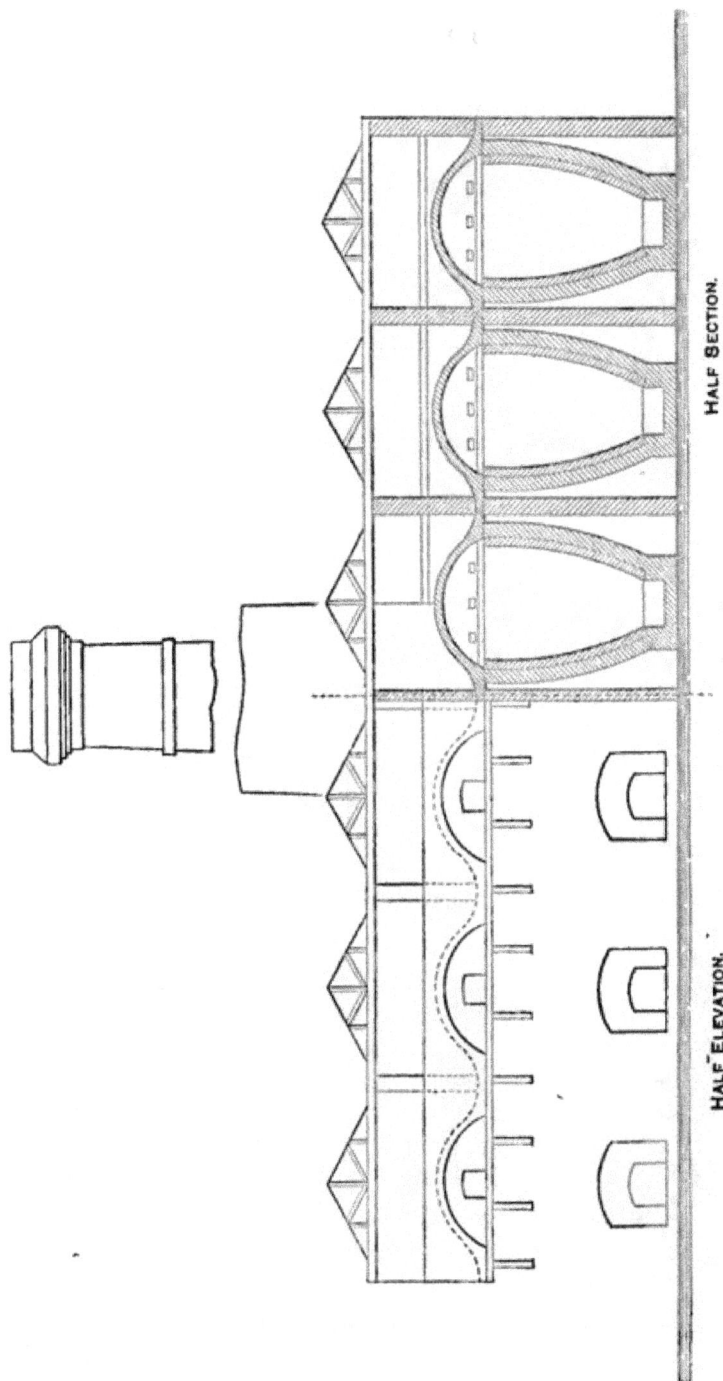

HALF SECTION.

HALF ELEVATION.

Fig. 5.—Michele's Patent Chambers.

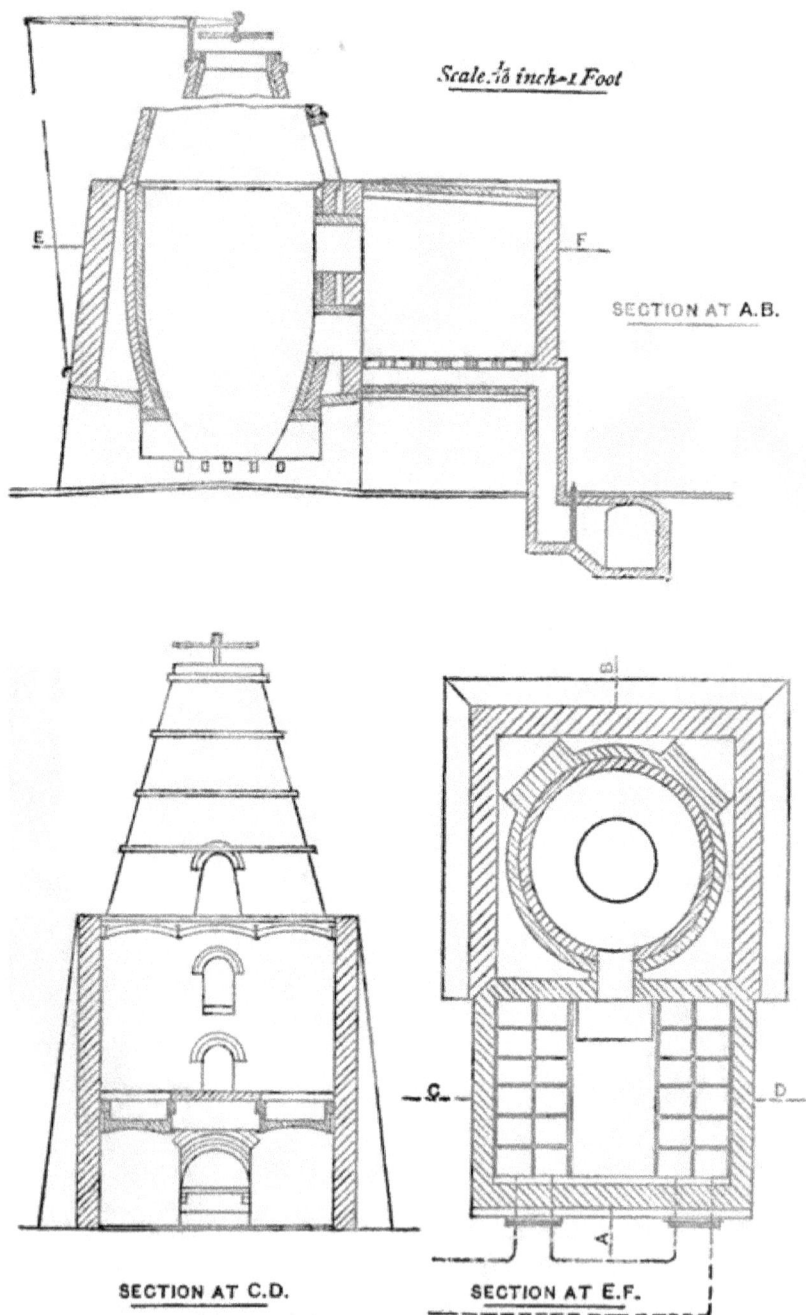

Scale. 16 inch = 1 Foot

SECTION AT A.B.

SECTION AT C.D.   SECTION AT E.F.

Fig. 6.—Spackman's Drying Chambers.

are drawn through the chamber, passing downwards in special
flues, constructed in the contents, to orifices in the floors of the
drying chamber, which will be seen in the section at A.B.   The
heat of the kiln in cooling is found in practice to be amply
sufficient to dry the contents of the chamber, which may conveniently be semi-dry bricks produced by means of any of the
machines at present employed for the manufacture of bricks of
this kind.   The size of the chamber is calculated to furnish a
full charge for the kiln, and there are suitable openings both for
filling the chamber and for loading the kiln.   It will be expedient
in arranging for the use of chambers of this kind, which are very
economical in working, to place the brick machines at such a
level that the bricks, as soon as they are pressed, can be wheeled
into the chamber and stacked in position.

Messrs. White's Drying Process.—Messrs. J. B. White &
Brothers attain much the same results by passing the surplus
heat from their kilns, which are somewhat small in size,
into long flues or chambers covered with removable iron plates.
These flues are so constructed that the kiln gases pass first
beneath the slip to be dried and then return above it into the
shaft, in their passage to which they heat an additional coating
of slip on iron plates.   The total thickness of dried slip thus
obtained is said to be from 15 to 16 inches for every square foot
of heating surface, and the total weight of fuel needed for drying
and calcining the slip is estimated to be 35 per cent. by weight
of the finished cement.

CHAPTER VIII.

THE CALCINATION OF THE CEMENT MIXTURE.

Cement Burning.—The burning of the cement is perhaps, of all stages in its manufacture, the one requiring the greatest care and attention, and it is the one which has received the least modification by the improvements which have been applied to cement making in recent years. It is true that many new forms of kilns have been proposed, and that the use of gaseous fuel and revolving cylindrical furnaces have been patented; moreover, numerous rather crude attempts have been made to save fuel, and to reduce the expense in this part of the process, by turning to account the waste kiln-gases, &c., but the original bottle-kilns, which have been employed since the first days of the manufacture, still hold their own, not only in England, but also on most parts of the Continent. The efforts made to use the Hoffmann and other fuel-saving kilns have not, so far, proved successful, at any rate in the hands of our English cement makers, and interstratified coke may be said to be the only fuel possible for the production of a good sample of Portland. Perhaps we should here except the Dietsch kiln, of which favourable accounts have reached us.

Chemistry of Cement Calcination.—We have already spoken of the chemical changes which have to be effected in the kiln—firstly, the expulsion of the carbonic acid gas, which is a matter of no difficulty; secondly, the production of compounds of silica, lime, and alumina at a point very little short of that of fusion. This is really the chief difficulty of the cement burner, and it is on this part of the process that our knowledge must be admitted to be lamentably defective.

It is almost impossible to say why in certain states of wind and weather this or that kiln, whatever the care taken in the loading, and however skilfully the fuel may be disposed and the draught adjusted, will in the one case contain a large amount of blue overburnt clinker, which the miller will be bound to reject, or in the other a mass of pink or yellow tender-burnt material, which will have to go back into the next kiln at a great waste of time and much expense.

Present Processes are Unscientific.—Our kilns, in the way we now use them, are certainly not scientific, but it is hard to indicate how they are to be improved. We are still in considerable ignorance concerning the exact degree of heat needed, the amount of atmospheric air required, the effect of the oxidising or the reducing flame, the beneficial action of steam, and many other matters which are at present rather too much in the hands of the cement-burner, and too little subject to exact scientific laws. As was well said in a recent article on the question of over-burning,* "it may be remarked that the dubiety that exists on this and other points in the practical manufacture of cement, which can only be dispelled by the accurate measurement of the temperature necessary for, and characteristic of the reactions it is sought to bring about, is an eloquent plea for the extension of the use of the pyrometer, devised by Le Châtelier, and introduced into this country by Roberts-Austen, to an industry as complex in principle as it is crude in practice."

Changes Effected in the Kiln.—To proceed, however, to the calcination, the chemist tells us that the lime on being freed from its carbonic acid gas, and, therefore, in what is termed the " nascent state," is in the best possible condition to enter into fresh combinations. As the temperature rises the draught in the kiln increases, and the sluggish and heavy carbonic acid gas is by degrees passed out of the kiln. The free lime is then presented to the silicate of alumina of the clay, and as laboratory experiments prove to us, it is enabled to expel the alumina and to form a fairly stable base, the calcic silicate, but in the proportions found by experiment to be those best adapted for the production of the Portland cement, there is invariably a much larger amount of lime present than the silica can neutralise. In order to account for the neutralisation of the lime, we are taught that the alumina, which was at first regarded as a base, assumes at very high temperatures acid functions, and enters into combination with a further portion of the lime to form the so-called aluminate of lime. The same may be true to some extent also with the iron.

Fremy's Experiments.—M. Fremy, whose experiments are relied upon for the existence of this alumina compound in Portland cement, was able to produce a base from lime and alumina alone in the absence of silica, and he thus dispelled the theory that the alumina still continued in combination with the silica in the form of the double silicate of lime and alumina.

In order to obtain the alumina in a perfectly pure state, M

* *The Engineer*, December 2nd, 1892, p. 473.

Fremy, whose researches were first published in 1865, prepared this substance by the calcination of ammonia alum. He also produced lime of absolute purity from Iceland spar. When this material is submitted to the intense heat of the wind furnace the resultant lime does not melt, but is transformed into a crystalline substance with a cleavage resembling that of marble.

Precautions taken to exclude Sulphur.—The experiments with lime and alumina were carried out in carbon crucibles, and to avoid the possibility of the formation of calcium sulphide, due to impurities in the fuel, the mixtures of alumina and lime were placed in a small inner carbon crucible, enclosed in an outer one packed full of powdered lime. In the utmost heat of the wind furnace, mixtures of 80 parts of lime and 20 of alumina were melted, as were also mixtures containing 90 parts of lime and 10 of alumina. The mixture of 93 parts of lime and 7 of alumina was even fritted or run to glass. The alumina, in fact, serves as an excellent flux for the lime and in this respect is superior even to silica. These compounds of lime and alumina, containing such a large excess of lime, are crystalline in structure, indeed their fracture resembles that of loaf sugar. They have a strongly alkaline reaction, and they combine with water with a considerable evolution of heat. They may in some respects compare with molten lime. These substances, which swell up in water, however, like quicklime, can obviously play no part in the set of cements, but the case is quite different when we have to deal with compounds of lime and alumina, represented by the formulæ Al_2O_3CaO, Al_2O_32CaO and Al_2O_3, $3CaO$, which are less basic than the foregoing salts. When these aluminates are reduced to a fine powder and gauged with water, they at once solidify and form hydrates, which acquire, even in water, a considerable degree of hardness; they are, moreover, capable of binding together large quantities of sand to form hydraulic mortars, which have a stone-like consistency.

Part taken by Alumina in the Setting Process.—As the outcome of these experiments, M. Fremy was led to assert that the set of cements was caused by the hydration of the aluminates of lime, and not by that of the compounds of lime and silica, which had no setting action, nor by the double silicates of lime and alumina, which in all different proportions he found to be likewise inert. If the accuracy of these investigations is admitted, we must relinquish all those theories of cement action which depend for their explanation on the behaviour of compounds of lime and silica.

Further Experiments by M. Fremy.—The setting action of the compounds of lime and alumina was, however, insufficient to explain all the phenomena of hydraulicity, as M. Fremy himself recognised, for it would not account for the behaviour of quicklime in the presence of puzzuolana or a calcined silicate of alumina. In a subsequent memoir in 1868 he showed that even in the case of a perfectly pure silicate of alumina, which contained 65 per cent. of silica, 25 per cent. of alumina, and 10 per cent. of water, and was represented by the formula $(SiO_3)_3Al_2O_3,2H_2O$, an excellent puzzuolana could be produced by simple calcination at a temperature of about 1,300° F. This substance, on the addition of water, was able to impart to lime hydraulic properties which were manifestly due to the simple dehydration of the clay, and it was found by further experiment that pure hydrated silicates of alumina, even when intensely heated, still retained their power of recombining with water and acting as puzzuolana, the only difference being that after exposure to very high temperatures the setting action of the mixture of lime and clay was retarded. The interval needed for induration could be extended or decreased in accordance with the temperature at which the silicate was calcined. He states his belief that all clays capable of acting as puzzuolanas, or such as contain relatively large quantities of silica which become gelatinous when treated with hydrochloric acid, are decomposed by dehydration ; that a certain proportion of the silica and alumina is not only decomposed, but these elements are separated and become extant in an allotropic form, and are thus ready to form new compounds when presented to quicklime in the presence of water. There are, however, as he points out, many puzzuolanas which contain also notable quantities of lime in addition to the silica and the alumina, and in these substances the effect of calcination is to bring about the formation of fresh compounds of lime and the silicates, in addition to the free silica and alumina, which substances are capable of rearrangement in the presence of water. He regards Portland cement then as a highly complex body, which must comprise aluminates of lime formed at extreme temperatures, but also notable proportions of these puzzuolana-like substances, all of them capable of becoming indurated in the presence of lime after treatment with water.

M. Le Châtelier's Experiments with the Microscope.— This theory, which is very ingenious, and not incapable of being reconciled with the views of Feichtinger and other chemists, received much confirmation from the more recent investigations of M. Le Châtelier. This latter savant has pursued his investigations by means of the polariscope. He finds on cutting thin sections of Portland cement clinker that the mass is composed undoubtedly of a variety of chemical compounds. It is quite impossible to isolate each of these small particles for analysis, but he has been able by synthesis to build up from lime, silica, and alumina, the different bodies which he recognises as forming part of the clinker, and he has ascertained that these substances, when viewed with polarised light in the microscope, are really the compounds he finds in the clinker.

Principal Substances Present in Cement.—The following are among the chief of the bodies present in Portland cement :—

1. A substance which has no action on polarised light. It consists of aluminate of lime, rich in lime, occasionally mixed with particles of free lime. M. Le Châtelier actually produced by artificial means a tricalcic aluminate, $Al_2O_3,3CaO$; he assured himself that this salt crystallises in the cubic, consequently polariscopically inactive, system, and that, moreover, it is the only possible compound of lime capable of being present in cements, besides lime itself, which crystallises in this system.

2. A substance acting feebly on polarised light and presenting a clearly defined crystalline form. This is a silicate of lime. M. Le Châtelier perceives in this the chief, if not the only active ingredient of cements ; this substance always constitutes the principal part and sometimes the entirety of Portland cements. He believes himself in a position to affirm that it is a calcareous peridot (lime olivine) $(2CaOSiO_2)$, which, upon the calcination of the cement, crystallises in the matrix described in the next paragraph, when the whole is carried to the point of fusion.

3. A substance having a deep brown colour which acts upon polarised light. This is the most fusible of the components of cements. It constitutes in the solid form the gangue of the silicate which has just been described, and when melted it is the cause of its crystallisation. It is an alumino-ferrite of lime, having less lime than the tricalcic aluminate. M. Le Châtelier assigns to it the formula $2(AlFe)_2O_3 3CaO$. ·

He was able to produce synthetically a compound answering to this formula, and found it to possess the optical characteristics and the ready fusibility of the similar substance present in cement. This material is slowly acted upon by water, and is but little changed during the setting process.

4. Small crystals having a very energetic action upon polarised light. These are by no means numerous, and undergo no change upon coming in contact with water. These are probably magnesium compounds, since M. Le Châtelier has ascertained that all very basic calcareous compounds are acted upon by water in contradistinction to those of magnesia.

Microscopical Investigations upon the Setting Process.—In examining the setting action of cements under the microscope by means of polarised light, M. Le Châtelier discovered that the addition of water leads to the formation of several different compounds. Of these the most important, as respects the induration, is a substance which crystallises in hexagonal laminæ, resembling the crystals of calcium hydrate, but he was unable to collect them in sufficient quantity to determine their composition. Whatever it may be, this material is undoubtedly a derivative of the calcium ortho-silicate. In quick-setting cements that are rich in alumina, there is a copious formation of long acicular crystals, which interlace in all directions. These crystals, when exposed to dry air, become dehydrated and shrink greatly in volume; if heated in water to a temperature of 50° C. they break up into fragments. They are produced by the action of water on tricalcic aluminate, which is slightly soluble in pure water, but more so in salt water, in which latter the crystals become partly decomposed. It may be the disruption of these crystals in certain defective qualities of Portland cement which causes the failure of the concrete in sea water.

The Causes of "Creeping" in Cement Clinker. — In the course of his experiments he discovered several other substances which did not act upon polarised light, but he was unable to find that they played any part in the setting process. He, however, ascertained that the calcium ortho-silicate beforementioned gave rise to the peculiar action of "creeping" often noticed during the process of cement manufacture. Heated to the point of fusion and then allowed to cool gradually, this substance at first took the form of a stone-like semi-translucent mass, which, however, crumbles into a powder, composed of fragments of minute twinned crystals, as it cools. This action is undoubtedly due to unequal tension on the opposite faces of the twins. If the compound is less intensely heated, the twins are not formed and no disruption takes place.

M. Landrin's Experiments on Cements.—Some important observations upon the silicate of lime have been communicated to the Académie des Sciences by M. Landrin, another patient investigator, who has done much to throw light upon this vexed question. He believes that in the compound of 44·55 parts of silica with 55·45 parts of lime, a salt which has the formula of $3SiO_2,5CaO$, we have the basis of all good Portland cements. He terms this substance "pouzzo-Portland," as existing alike in Portland cement and puzzuolana. It would seem from his experiments that it constitutes a large proportion of all cements

of the Portland type, along with aluminates of lime, oxide of
iron, and magnesia. It has been observed that when equal
quantities of lime and silica, in a pure state, are strongly heated,
a compound is obtained resembling Wollastonite, with the
formula $CaSi_2O_3$, which contains 48·4 of lime and 51·4 of silica.
This substance, which is very hard and crystalline in fracture,
is not attacked by water, and thus is not capable in any
way of influencing cement action. If the volume of lime is
doubled so as to obtain a compound answering to the formula
Ca_2SiO_4, containing 65·1 per cent. of lime and 34·9 per cent. of
silica, the resulting silicate is when hot a hard compact mass,
but as it cools it gradually disintegrates and crumbles away into
a bulky grey powder rather resembling slaked lime. This is no
doubt the substance, also termed the ortho-silicate, which is
present in so-called "slippy clinker," produced from over-clayed
slurry.

   The Chemistry of the Subject still uncertain.—All
these theories and observations, though some of them tend to
throw light upon difficulties encountered in practice, do not to
our mind thoroughly explain the phenomena observed in cement
action, and we must leave this branch of the subject with the
conviction that we have still much to learn concerning what
really takes place both in the kiln and on the subsequent addition of water. There can be no doubt, we think, that in the
intense heat of incipient fusion the iron also acquires acid properties, and forms more or less stable compounds with the lime
and the magnesia.

   Cement Equations (Bases in Proportion to Acids).—Some
observers have attempted to form equations for the acids and
bases in any given sample of cement, thus—

$$\frac{CaO + MgO}{SiO_2 + Al_2O_3} = 2 \cdot 45 ;$$

but in the absence of the knowledge of the exact behaviour of
the alumina and the iron, whether as acids or bases, we are
unable to found any very trustworthy theories on this and
similar equations. We may discard the alkalies from all consideration, for it is no doubt true that in well-burnt cement
samples they are to a large extent volatilised and dispersed in
the kiln. Indeed, the presence of a notable proportion of potash
and soda certainly points to under-burning. The safety limit of
lime in the slurry depends in a very high degree upon the
temperature at which the calcination is conducted, and the black
dense granular clinker produced in a gas-flame in the revolving
cylinder of Mr. Ransome may apparently contain as much as

65 per cent. of lime, as will be seen from Mr. Spackman's analysis
of a sample of Portland cement prepared in this way at the
South Wales Portland Cement Works, Penarth, near Cardiff.

Insoluble residue,	2·368
Silica,	18·433
Ferric oxide,	4·453
Alumina,	7·235
Lime,	65·446
Sulphate of lime,	·162
Magnesia,	1·913
Potash,	·076
Soda,	·117
	100·203

The extraordinarily small percentage of lime in the cement
produced from blast furnace slag, which has many of the properties of Portland cement, is, as we shall explain later, very
subversive of the generally accepted equation for the amount of
acids and bases.

Fuel used in Cement Burning.—It is almost impossible to
lay down any exact rules for the time occupied in cement
burning, and the amount of fuel employed varies also considerably. The results with some carefully-weighed kiln charges,
burnt in the ordinary way with interstratified gas coke at the
Folkestone Cement Works, as communicated to us by Mr.
Charles Spackman, F.C.S., were as follows :—The slurry, which
was specially well dried for these trials, was found on the mean
of several determinations to contain 2 per cent. of water only.
The coke also was fairly dry. Pieces of moderate size were
used, except in the bottoms, for which larger lumps were picked
out. Five brushwood faggots were used in each case to start
the fires. In kilns Nos. 3 and 7 all the materials were accurately
weighed in and the product was weighed out.

The weight of coke was in No. 3 kiln—

	Tons.	cwts.	qrs.	lbs.
In bottom on faggots,	0	7	1	24
,, layers interstratified,	7	4	0	20
On top of kiln,	0	7	0	0
Total,	7	18	2	16

This kiln yielded 19 tons 2 cwts. 2 qrs. of good clinker and
6 cwt. 3 qrs. of yellow, and, therefore, to produce 1 ton of clinker
the coke required was 8 cwts. 1 qr. 5 lbs. As the mean of
repeated observations 50 bushels of gas coke fresh from the

6

works weigh 21 cwts., and, therefore, 8 cwts. 1 qr. 5 lbs. are equal approximately to $19\frac{3}{4}$ bushels. The coke cost $2\frac{1}{2}$d. per bushel, and, therefore, the cost of the coke for burning 1 ton of cement was almost exactly 4s.

In No. 7 kiln, with the same careful observations, 7 tons 17 cwts. 2 qrs. 16 lbs. of coke produced 19 tons 7 cwts. 2 qrs. of clinker and 2 cwts. 1 qr. 12 lbs. of yellow—the coke used per ton of clinker was thus 8 cwts. 15 lbs.; say 19·36 bushels, costing about 4s. as before.

In a third trial, the weights in which were arrived at by an average, each ton of clinker needed 18·66 bushels of coke, or say 3s. 11d. per ton for the fuel for burning.

Fuel used with Johnson's Chambers.—As compared with these experiments, we may place the facts kindly communicated by Mr. Johnson on this subject. He states that the cost of fuel for the drying of the slip into slurry and used in the kilns on his system amounts to 5s. 3d. per ton of the finished cement. If to Mr. Spackman's figures we add 1s. per ton for the cost of fuel required to dry the slurry, we obtain an estimate which compares fairly well with that of Mr. Johnson. But Mr. Johnson's totals, based on the working under the old system of kilns of large size, partly fired with gas coke, but in the main (to the extent of two-thirds) with coke made in ovens beneath the drying floors, show a much larger cost. He estimates the expense of the gas coke and oven coke for burning 133 tons of cement clinker as follows :—

50 chaldrons of coke-oven coke, at 16s. 10d. per chaldron,	£42	1	8
25 chaldrons of gas coke at 10s. 6d. per chaldron, . .	13	2	6
	£55	4	2

Or the cost per ton equals nearly 8s. 4d. Of course, this includes also the cost of drying the slip over the coke-ovens.

Calcination in Ransome's Cylinders.—Among the more scientific attempts to improve the calcination of Portland cement, we must undoubtedly place the process of cylinder burning, invented by the late Mr. F. Ransome, the description of which we have obtained from the paper read by him before the Mechanical Science Section of the British Association at Manchester in 1887. He substitutes for the somewhat costly and wasteful kiln a revolving furnace, cylindrical in form, constructed of an outer casing of steel or boiler plate, lined with good refractory fire-bricks, so arranged that certain courses are set forward in order to form three or more longitudinal projecting fins or ledges. The cylinder is rotated slowly by means of a worm and wheel, driven by a pulley upon the shaft carrying the worm. The cylindrical casing is surrounded with two circular rails or pathways, turned perfectly true, to revolve upon steel rollers mounted on suitable brickwork. It is provided with regenerative flues, by passing through which the gas and air severally become heated before they meet in the combustion chamber at the mouth of the revolving furnace. The gas may be prepared from small coal or other hydrocarbon, burned in any suitable gas-producers (such, for instance, as those patented by Messrs. Brook & Wilson, of Middlesbrough, or by Mr. Thwaite, of Liverpool). The producer may be situated in any convenient spot in the vicinity of the furnace.

Raw Materials Used in the Form of Powder.—In lieu of employing a liquid mixture of lime and clay dried into lumps, the slurry must be dried and crushed between rollers into the form of a powder. This powder is conveyed from the crusher by means of an elevator, and it is discharged into a hopper from which a regulated amount drops into the revolving cylinder. This cylinder is set at an angle of about 10° or 12° with the horizon, and, in the act of turning, the charge of cement material gradually travels from the upper to the lower or hot end of the furnace. In so doing it falls a great number of times across the hot flame, the cylinder being in motion lifts the powder on its advancing periphery, where it rests against one of the projecting fins or ledges until it has reached such an angle that it shoots off in a shower through the flame, and falls once more on the lower side, to be again lifted by the revolution of the cylinder, and again dropped through the flame. This action takes place at each complete revolution of the cylinder, and in accordance with the angle of inclination the number of times the charge is turned over may be made greater or less. By the time the powder has reached the lower or hotter end it has pursued a roughly helical path, during which it has been lifted and shot through the flame a large number of times, and its passage may have occupied about half an hour.

To some who have been accustomed to the protracted and tedious process of kiln-burning, the time thus occupied in the revolving cylinder may appear insufficient to effect the combinations necessary to convert the raw material into finished clinker, but on careful consideration it will be evident that the conditions which here prevail are precisely those best adapted to bring about in the most perfect manner the physical and chemical changes needed for the production of Portland cement. As the raw material is in powder it offers every facility for the speedy liberation of the water and the carbonic acid gas, and this operation is no doubt accelerated by the velocity of the current of burning gases through which the particles of powder have to pass. Again, the fineness of the particles results in their being speedily heated to a uniform temperature, so that they do not serve as nuclei for the condensation of the moisture expelled by the heated gases.

Economy of Time by this Process.—On reaching the lower end of the furnace the gases pass into the exhaust-flue, and the calcined material is discharged on to the floor, or on to a suitable "conveyor," and removed to a place of storage where it may be cooled and subsequently ground and finished. The material in lieu of being in the condition of tough clinker, requiring to be broken up before passing into the mill, is in the state of a coarse granular powder, somewhat resembling gunpowder. The operation of burning is continuous, as the furnace when once started goes on day and night, receiving the adjusted quantity of powdered material at the upper end, and discharging its equivalent in dense well-burnt clinker at the exit end. A week's work in the case of a kiln is thus done in one day with the cylinder.

Yield of Cement per Cylinder.—The saving thus effected in cost of fuel and wear and tear to brickwork of kilns is doubtless very great. Such a furnace as above described is capable, it is said, of turning out 20 tons of cement in 24 hours at a consumption, so it is stated, of 3 tons of small coal. There is also a great economy of space as well as a saving in time.

Our own experience of the working of this process does not, however, justify the above description, which would lead one to believe that the system is almost theoretically perfect. There is a great tendency for the powdery clinker to ball together in the cylinder, due to the incipient fusion of the silicates. In order to avoid this it is necessary to keep the percentage of lime very high, and the result is a treacherous overlimed cement. We are not aware that in any case this invention has answered the expectations of manufacturers, though we should like to see some such process substituted for the wasteful system of calcination in general use.

Stokes' Cylinder-Firing Process. — Another inventor, namely, Mr. W. Stokes, Assoc. M.I.C.E., has busied himself with cylinder-firing, and he employs the surplus heat from the furnace for drying the slurry. The apparatus is so arranged that slurry, prepared by the semi-wet process, is pumped in definite quantities into a long trough, so contrived that a revolving drying-drum, placed above it, dips into it. The waste products from the furnace pass through this drum and raise it to such a temperature that the coating of wet slurry, which adheres to its external surface, and is carried round with it, is dried during the course of a single revolution of the drum. The caked slurry is removed by means of a scraper, which strips it off the surface before the cylinder again dips into the trough or tank. The exterior surface of the drum is surrounded with rings or gills, which increase the drying action, and are filled up with the plastic slurry in the trough. The dried slurry, in the form of small sticks or fingers, falls into a creeper, by means of which it is fed into the upper end of an inclined revolving furnace. It slowly passes through the firing cylinder, being turned over repeatedly during its progress. The firing is effected by the flame of gas-producer gas, for the manufacture of which due provision is made. The gas enters at the lower end of the cylinder, and there encounters a current of heated air, which issues from a revolving cooling-chamber placed beneath the firing cylinder. This chamber, which is also cylindrical in form, receives the finished clinker as it issues from the furnace ; and the waste heat, given up by the clinker in the process of cooling, serves to heat the air to a temperature of about 900° F. The furnace-chamber makes 1 revolution per minute and the cooling chamber $1\frac{1}{2}$ revolutions in the same time. This process, which appears to be thoroughly sound and scientific in principle, has, we understand, been successfully applied on a manufacturing scale.

Advantages of this Process.—One of the great advantages no doubt of this and similar processes is the freedom from admixture with the ash of the fuel, while the absence of deleterious fumes would point to the adoption of this plan of burning in places where cement works have caused a nuisance (see Appendix B, p. 207). The cost of labour is, moreover, very small as compared with the present plan.

Other Systems of using Gaseous Fuel.—It is easy to conceive other systems by which gaseous fuel might be employed without the mechanical difficulties entailed by the revolving cylinder and the liability of the cement mixture to ball together and clog. Thus General Scott proposed a vertical shaft through which the powdered materials were allowed to drop by the operation of a series of movable ledges or buffer plates, the calcination being accomplished by means of gaseous fuel.

Admixture of Ash in some Cases of Benefit.—A recent writer on this subject considers that the ash of the fuel in an ordinary interstratified kiln may, under certain conditions, prove a " pure boon to the careless and incompetent manufacturer," as he infers that it may have " a most undesirable levelling-up tendency," and he proceeds to show that in cases which may possibly arise, the ash from the coke may be equivalent to 5 per cent. of the weight of the cement, and may add such a percentage of silicates as to change the character of the resulting compound. He states that the time is ripe for the advent of the gas-producer as a necessary part of a cement-making plant.

CHAPTER IX.

THE GRINDING OF THE CEMENT.

Milling Machinery a much disputed Subject. — The nature of the milling apparatus to be used for the reduction of the cement to a fine powder has long been a subject of grave contention among practical men, but we believe we are justified in stating that, at anyrate in this country, after many trials of much vaunted machines which promised to do the work better and more cheaply than the old-fashioned mill-stones, used for grinding wheat into flour, cement makers have, in the long run, been obliged to confess that nothing could beat the performance of well-dressed "French burrs."

Character of the Meal Important.—The reason of this preference is not far to seek, for there can be but little doubt that the character of the meal produced is of much importance and greatly affects the quality of the cement.

The ultimate particles of a good sample of Portland must present under the microscope the appearance of little sharp splinters rather than of rounded sand-grains, and the reason why mill-stones turn out such good work is that they rub down the hard particles of clinker into sharp fragments rather than crush them into cube-shaped atoms or roll them into rounded granules, such as those produced from edge-runners and roller-mills.

Crushing Machinery.—Before passing on to the grinding machinery of various types, we may state that both the cement clinker and the raw materials for treatment by the dry process require crushing, which is now usually done by stonebreakers. Previous to the general introduction of these machines, crushing rolls and edge-runner mills were employed for this purpose, and in some cases those appliances are still in use. In a recently reconstructed manufactory edge-runners have been employed, and the product thus obtained is sifted; all the core rejected by the sieves is ground by millstones. It is said that, after crushing, 59 per cent. will pass a 60 by 60 mesh sieve.

To recommend any particular make of stonebreaker would be out of place here. Most of these machines are modifications of the original patent of Blake ; they may be obtained of all sizes, and every maker of any repute can give references to cement manufacturers using his machines. The only points requiring notice are the sizes to be selected and the method of fixing them. For clinker the size may be regulated by the quantity required to be ground per hour, the smaller machines being, as a rule, of sufficient strength. It is always advisable, however, to select one of larger capacity than is absolutely needed for the hourly grinding, so as to keep a constant supply in the storage-bins for use at meal-times and when grinding at night.

Size of Stonebreakers to be Used.—For dry process raw materials the matter must be looked at in a different way. Many of the limestones in use are extremely hard, even some of the lias beds are very tough and difficult to break. If only a small output is required it would be unadvisable to select a machine rated to break this quantity only, unless there is ample evidence that it is of sufficient strength. It is really not a question of how much the machine will break per hour, but whether it is strong enough to break the material at all. For crushing these substances the stonebreaker is undoubtedly the most efficient machine, and in small works, if the stonebreaker is of larger capacity than is required for the raw materials, it may, by a system of conveyors or shoots, be also used for the cement clinker. It will, we think, be found an advantage where a very hard material has to be broken to a very small gauge, to do the work in two machines of different sizes—a large, powerful machine for the first break and two or more small machines for the reduction of the coarsely-broken material from the larger machine.

Mode of Fixing Stonebreaker.—The stonebreaker should be fixed on a timber foundation, the mouth being on a level with the floor, so that the contents of a barrow may readily be shot in ; it should have an iron shoot leading from beneath the jaws to the elevator fixed at either side. Ample space should be allowed at the sides to enable the flywheels to be taken off, and there should be plenty of room at the back as also underneath the machines for getting at the drawback rods and toggle-plates. Beams with ring-bolts for lifting tackle should be fixed overhead for use when replacing broken parts, or when attending to the bearings.

Elevators from Stonebreaker.—The elevator to the storage bins may with advantage be of steel-link chain, many forms of which are in use, or Greening's wire belt may be employed, the latter being used on ordinary pulleys. Stamped steel buckets are rapidly replacing the older forms. These are of sheet steel blocked from the plate in one piece, without seams, rivets, or overlapping edges. They are made in all sizes up to $\frac{3}{8}$ inch thick, and of several patterns, to suit various requirements, and for use either with vertical or with inclined elevators.

Alternative **Methods of Crushing.**—Among the numerous machines now before the public, which have been introduced for the purpose of crushing and pulverising the raw material or the clinker for cement-making, we have selected a few for special reference, either because they represent characteristic types of the plant in question, or because they have been successfully used in cement works which have come under our notice.

The Lowry Crusher.—The crushing action of the pestle and mortar, when the pestle is simply rotated in the mortar and not in the form in which it is used as a stamp, has been taken as the model by more than one inventor for stonebreaking as also for grinding purpose. For the former use the machine of Lowry seems well designed. It is called the "continuous gyratory-action stonebreaker," and has the merit of compactness, simplicity of construction, and extraordinary crushing power. It is not subject to the strain prevalent in machines with flat crushing-faces, and the circular shape causes the stone or clinker to bear on its outer edge, leaving a hollow behind where the impact or blow is given, thus ensuring a proper distribution and consequent reduction of strain. The shaft does not revolve or vibrate, but gyrates, and by means of this gyratory motion and the circular form of the crushing surfaces, the principal feature—i.e., the cracking action of the machine—is obtained, and the stone is broken without the tendency to split or flake, as in all descriptions of jaw-crushers. Although the wearing parts are few and the frame-work is strong, the machine is designed with all due consideration to future repairs, and every part is easily accessible and can be readily duplicated or renewed as time and circumstances may require. The machine while in motion may be adjusted to break to any required gauge.

Work done by Revolving Crushers.—In comparing the gyratory crusher with the older form of these machines, it is clear that the efficiency or output will depend upon the quantity of the material which can be operated upon in any given period of time, and it is evident that that type of machine will be most efficient which brings into action the largest effective crushing surface in the time in question. Taking a well-known jaw-crusher, for the sake of comparison, and a No. 4 Lowry breaker (size 10" × 20" having a circular crushing head, with an external surface of 1,303 square inches, and a corresponding annular crushing-block with an internal surface of 1,971 square inches) we have a total crushing surface of 3,274 square inches, and if this machine is driven at 120 gyrations per minute we obtain, for the material in contact, an actual crushing surface of 392,880 square inches per minute. A jaw-crusher, taking stones of the same size, with a square head 16″ × 18″, and a corresponding square crusher-block of 16″ × 18″, has a total crushing surface of 576 square inches, and if the machine makes 250 strokes per minute it will bring into contact an actual crushing surface of 144,000 square inches per minute; the proportionate amount of surface being thus nearly 3 to 1 in favour of Lowry's gyratory machine.

Mode of Feeding the Lowry Crusher.—Another advantage connected with this machine is that it does not require hand-feeding, but the substance to be broken may be tipped into the hopper by the cartload at a time. In some trials of this system, a ton of hard slag was broken to macadam size as employed for ordinary roads in eight minutes and a second ton in five minutes.

Multiple Disintegrator.—Among the most recent improvements of machines of the well-known disintegrator type is the New Patent "Multiple" Disintegrator, manufactured by the

Fig. 7.—Patent Multiple Disintegrator.

Hardy Patent Pick Company, Limited, of Sheffield. The feature of this machine is the employment of a series of beaters small in circumference at first and gradually increasing in size as the work of reduction progresses. Each set of beaters works in a separate chamber, the whole circumference of which is provided with serrated linings of excessively hard chilled iron. These linings serve as grinding surfaces, and it is claimed that by this arrangement there is here about eight times the area of grinding surface to be found in any other percussive disintegrator of equal size. Moreover, as the chambers increase

in diameter as they approach the outlet-end, the fan action
of the larger beater draws in the air with great velocity and
produces a strong current passing from the inlet- to the outlet-
end of the machine. The action of this air-current tends to
carry away the finished material as soon as it is reduced to
a sufficiently fine powder. The substance to be reduced, as will
be understood from the accompanying illustration (Fig. 7), enters
the smallest beating chamber, and from thence passes in succession through all the others, being subjected during its transit to
continuous percussive action, increasing in intensity in each
chamber of the series. It will be understood from this description that the machine is not dependent upon grids or screens
for the production of fine powders, as is the case in many
machines of this kind, but provision is made for the insertion
of a screen near the outlet, should it be considered advisable to
employ one.

Special Advantages of this Machine.—The special features
of this disintegrator are its large grinding surface, its adapta-
bility for fine grinding, the production of a beautifully soft and
even sample, and the treatment of materials which could not
otherwise be dealt with because they would instantly clog up any
other percussion machine of this type. It is stated that even
moist substances are capable of reduction to powder. The wearing
parts are of the simplest construction, and are consequently
cheap and easily replaced. These disintegrators are made in
three sizes, the smallest being capable of turning out from 2 to
4 cwts. per hour and weighing only 6 cwts., the largest being
designed to produce from 8 to 12 cwts. of fine powder per hour,
the weight in this case being 40 cwts., and the approximate
N.H.P. for driving them 12 to 15.

Grinding with French Burr-Stones.—Though we shall
have to notice in this section of our subject some of the many
contrivances that have recently been brought forward for cement
grinding, we prefer to bring into chief prominence the system of
grinding by French burr-millstones. This method is the one
most largely used; indeed, it is employed almost universally in
the Thames and Medway districts, where the mills in their present state of efficiency have been evolved from those once so
largely used for grinding corn.

Early Method of Driving in Use in certain Works.—Some
of the early works were fitted with spur-gear, and in at least one
case this system is still in use. The flywheel-shaft of the engine
was carried through the engine-house wall, and by means of
bevel-wheels it was made to drive a vertical shaft, on which was
fixed overhead a large spur-wheel, round which the mill-stone·

spindles were arranged and driven off it by pinions. The spur-wheel and one of the bevel-wheels were geared with wood. This system was anything but a satisfactory one, as in case of any accident to either of the wood-geared wheels (a by no means uncommon occurrence), or when the necessity arose for re-gearing either of them, the whole works would come to a standstill.

Improved Method of Driving.—An improvement on this plan was the system, now always adopted, of driving each pair of stones from a horizontal or lay-shaft by means of bevel-wheels. One end of the shaft was driven either by a spur-wheel fixed on the flywheel shaft of the engine, or by a spur-flywheel. The required speed of the mill-stone spindles was got up gradually from the engine, the lay-shaft making perhaps 80 to 90 revolutions per minute, the additional speed required by the spindles being obtained by the bevel-wheels. Here any failure of the bevel-wheels could be quickly remedied by the use of spare ones, but, as in the case of the spur-gearing, any accident to the main pair of wheels stopped the mill. By this arrangement, however, the risk of total breakdown is not nearly so great as with the horizontal spur-wheel, because there is only the one set of main gearing instead of two. Many mills of this kind are still at work. On visiting a large cement factory a year or two ago, we found it was driven by a beam engine which had then been standing for three weeks owing, as our guide said, "to a segment being out of the wheel."

Modern Arrangements for Driving Mill-Gear.—With the introduction of horizontal engines, running at higher speeds than some of the early beam engines, the flywheel shaft was in many cases coupled direct to the mill-shaft, the required speed for the spindles being obtained by the bevel-wheels. By this method the risk of a general breakdown is much reduced; the drawback is the wear of the wheels. The pinions on the spindles are necessarily of small diameter, and if both wheels are of metal the pinions wear out very rapidly, while, even if the wheels on the shaft are geared with wood, their life is not of very long duration. Any shock or strain in the mill is, moreover, transmitted to the engine, which should always have a heavier flywheel than would be necessary if the power was taken from it by any of the usual methods. This arrangement has often been adopted, with more or less success, with engines running at from 60 to 90 revolutions; it is most satisfactory either with a pair of coupled single engines or a coupled compound engine.

Rope Driving Advocated.—All things considered, the best method of driving is by ropes, although the loss by friction is with ropes greater than by gearing, and the ropes in time require renewing. The first cost of either plant would be about the same, as, on the one hand, we have to set the extra cost of flywheel by preparing it to act as a rope-drum, together with ropes, drum on mill-shaft, and additional length of building, against the cost of gearing with its bearing foundations. On the whole, we think the loss by friction and the occasional cost of renewing ropes is amply compensated for by immunity from loss by breakdown, to which gearing is always liable. For each pair of 4 feet 6 inch stones used for grinding clinker 40 I.H.P. should be allowed, and for slurry or the raw material of the dry process 30 I.H.P.

Cement - grinding Machinery must be Powerful.—Although the general principle of grinding is the same in both cases, mills for cement require to be of much stronger construction than those employed for corn, or indeed for most other substances usually ground by mill-stones. This is self-evident, when it is considered that the mill-stones themselves are heavier, that the material to be ground is of great hardness, that a much larger output is required, and that allowance must be made for pieces of iron, such as bolts or nuts, tools, &c., getting into the mills. This is a matter of not infrequent occurrence, due largely to the carelessness of workmen, and with weak gearing much damage may sometimes be done before the mill can be stopped and the obstruction removed.

Ordinary Rules for Flour Mills of Little Use for Cement.—The ordinary rules adopted by mill-wrights for the calculation of strengths are of little use for designing cement-mill machinery, and success in the past has been due more than anything else to the experience gained by failure, and by the system of trial and error

The Hurst and its Construction.—The "hurst" is the cast iron framing carrying the pedestals for the lay-shaft, the bridges for the footstep-bearing of spindle, and the lightening gear for raising and lowering the upper stone or runner. The hurst for each pair of stones is complete in itself, and consists of a bottom casting or bedplate, on which four columns rest, which support a pan to contain the bedstone. This pan has four screws passing through the bottom of it, on which the bedstone rests, and by means of these screws the stone can be raised and levelled as it wears away. It has also four other screws at the sides, by means of which the stone can be adjusted laterally. The hurst is securely bolted down to a concrete or masonry foundation, and any required number of hursts can be bolted to each other, thus forming a structure of great rigidity, by which the whole of the machinery is carried. The mill- or stones-floor is laid level with the tops of pans, the columns being of sufficient height to allow head room underneath.

The Lay-Shaft.—The shaft should be swelled at the places where the wheels are put on and at the driving end for a rope drum. The pedestals are bolted to the bedplate castings. Sometimes one-half of the number of mills are arranged to drive in the opposite direction to the other half, with the object of equalising the strain on the shaft.

Diameter and Speed of Stones.—For grinding, either the slurry of the wet process, the raw materials of the dry process, or the cement clinker, mill-stones of 4 feet 6 inches diameter are most suitable. For slurry the speed of these may be as high as from 160 to 170 revolutions per minute, in the other cases 140 revolutions will be found the best. When the mill is rope-driven the spindles may of course be driven by mitre-wheels. The extreme accuracy now attained in cast wheels by the use of modern moulding machines has rendered the wooden geared wheels almost obsolete, but at the same time there is much to be said in their favour for mill-stone driving. They run with less noise than iron wheels, but the first cost is greater; the iron wheel, however, never wears out. If they have been properly designed and are of sufficient strength, the wood-geared wheels will last a long time before the teeth require renewal.

The Geared Wheels.—If wood wheels are used, the one on the shaft should be the geared one, and it should be made in two halves, a spare one being kept in case of renewal. If both wheels are of iron (or steel), the first cost will be less, but they will in time both wear out, and one or both may at any time be rendered entirely useless by accident. The one on the shaft should also be put on in halves to facilitate renewal.

Footstep-Bearings and Neck-box.—The footstep-bearings of the mill-stone spindles are carried by the bridge, the lightening gear being actuated from the stones-floor, so that the grinding can be regulated without compelling the miller to descend to the room below. The upper end of the spindle is carried by the neck-box, which is sometimes cast upon the stone-pan, and sometimes securely wedged into the bedstone itself. Objections have been made to the latter method on the ground that the box is liable to drop out; if, however, it is, in the first place, properly secured with dry fir wedges there is not the slightest fear of this. The old style of box with a flange resting on a rebate sunk in the stone and run in with lead is a great source of trouble, as it has to be taken out in order to let the flange in deeper as the stone wears, while the box that is fastened in with wedges can be driven down from time to time, and, if necessary, tightened by driving a few flat iron wedges into the wooden ones.

The Brasses and the Method of fixing the same.—The brasses fit into recesses in the box but do not completely encircle the spindle. Recesses are also left where the brasses do not occur, and these are packed either with hemp or with clean cotton waste and suet, well tamped in. The whole bearing is covered by the usual dust-plate and cap, and if care is taken no dust can enter the neck, which will be sufficiently lubricated until the stone requires dressing, when the packing can be renewed.

The Runner-Stone and the Feeding Arrangements.—The method of casing the runner-stone and the feeding arrangements are in principle the same as in the case of corn mills, and are sufficiently familiar, except that case, shoots, &c., should be made of iron or steel plate $\frac{1}{8}$ inch thick ; the use of wood for this purpose being now entirely abandoned.

Wilder's Self-acting Feed.— A recent improvement is Wilder's self-acting feed, by which the material to be ground is fed into the eye of the stone by a screw instead of the damsel and shaking-shoe. The stone-case should be provided with an angle-iron ring, riveted round the lower edge, for securing the same to the floor, and there should be an opening closed by a slide near the floor and over the spout in order to test the grinding.

Mode of Building up Stones.—The mill-stones are built of specially selected French burrs, and are jointed and backed with Portland cement. The runner when new will weigh about 30 cwts., and as it wears it should be occasionally rebacked with cement and pieces of flat iron, to bring it back to somewhat near the original weight. Two of the eye-burrs in the runner should be of greater depth than the others to support the cross-bar. As this bar has to be let in deeper as the stones wear, it is best to cut the incisions for it to the full depth when the stones are built, and to fill them up again with Portland cement. The cross-bar, which is secured by being run in with lead, can thus be let in without loss of time as the cement of course cuts away much more easily than the French burr-stone would do.

The Method of Dressing the Stones.—The dress of mill-stones is different to that used for grinding corn. A plain skirt from 6 to 7 inches wide, without any furrows whatever, is left on both bedstone and runner. These surfaces are dressed with a pointed bill, and they should be rough but perfectly true. The surface of the bedstone within the skirt should be kept level with the skirt or slightly below it, by the occasional use of the hammer and pritchel, and may be left rough—the rougher the better. The runner is cut away from eye to skirt, as shown in sketch (Fig. 8), this part being called the "swallow," and like the inner part of bedstone, it should, in a similar way, be kept rough. About six trough-shaped furrows 7 or 8 inches long should run inwards from the inner edge of the skirt on both bedstone and runner, and between them there should be smaller cracks or furrows about 2 inches apart. These furrows are not in a radial line, but should make a slight angle with a radius in the direction in which the stones run, and their size depends on the gauge to which the material to be ground has been broken.

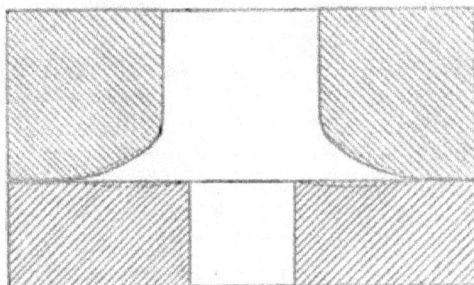

Fig. 8.—Cross-section through French Burr-stones.

Minor Details in Mode of Dress best left to the Miller.—This is the general principle on which the stones are dressed, either for grinding slurry, the raw materials for the dry process, or the cement clinker; but every cement miller has his own ideas as to the width of skirt, the angle, and the shape of the furrows, &c., and if a qualified man he may usually be left to himself in matters of detail.

Skirts should be Narrow, to avoid Excessive Friction.—As a general rule, however, it is advisable to work with as narrow a skirt as possible, as the wider it is the more power is required to drive the stones. The inner hoops should be kept level with the faces of stones, to prevent as much as possible the outer edge of the skirt from breaking away.

It is impossible, however, altogether to prevent this, and the joints in skirts, if broken away, as also any damage to the outer edges, should be carefully run up with lead whenever this becomes necessary.

Mode of Feeding Stones.—The material to be ground is fed regularly into the eye of the runner. As the stone revolves it is carried towards the circumference, and by friction against itself, together with the shearing action of the furrows, it is gradually reduced in size until it reaches the skirt, on which the real work of flouring is done.

System of Conveying away the Ground Cement.—From the mill-stones the cement falls into spouts, which discharge either upon a belt or into a trough containing a screw-conveyor, by either of which means it is removed to the sifting machinery or to the store, if sifting is not practised. On the whole, we prefer the screw-conveyor, which has been greatly improved during the past few years. The old cast-iron worm has been quite discarded, and has been almost universally replaced by the well known spiral form of skeleton thread or blade, which may be obtained to carry any required quantity, and which is supplied with suitable steel troughing. Care should be taken that the conveyor is not only of ample capacity for the quantity to be dealt with, but also that it is of sufficient strength.

In many factories sifting is not practised, the common specification of 10 per cent. residue on a 50 × 50 mesh sieve being met by careful grinding. If a finer product is required sifting, if not absolutely necessary, is always advisable.

Cement Sifting.—Sieves may be divided into two classes, rocking or shaking, and revolving sieves, both of these plans admitting of many modifications. With the former kind it is the usual and best arrangement to make each sieve deal with the product from one pair of stones ; one or more pairs of stones may discharge into a revolving sieve. Revolving sieves may be either hexagonal or octagonal in cross section, 2 feet being a suitable width for each of the sides. The length of the sieve itself is, of course, governed by the quantity of material required to be passed through it and by the necessary fineness. The wire cloth should be of steel wire, which is now specially woven for sifting cement. The cloth should be strained on very tight, as tight as a drum head, and it is best to attach it to removable frames, which can be kept strained in readiness to replace damaged ones. Suitable arrangements should be provided for keeping these sieves in a constant state of vibration.

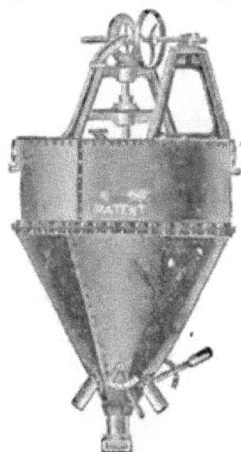

Fig. 9.—Askham's Separator.

Separators for Cement Sifting.—Separators, by which the coarse and fine particles are separated by the action of a current of air, have during the past few years to some extent replaced sieves in cement manufactories, with more or less success, that known as Mumford and Moodie's Patent, which is manufactured by Messrs. Askham Brothers & Wilson, of Sheffield,

7

and which was first applied to mills for grinding phosphates, is the one most generally used. Its general form will be seen from the illustration on p. 97; it consists of an inverted cone of sheet-iron, and an inner cone having a cavity between them. The ground material is fed by a hopper into the inner cone and falls upon a revolving disc. As it is thrown from this it encounters a current of air supplied by a fan, the effect being that the finer portion is carried over into the cavity between the cones and is disharged by a spout at the bottom, the coarser portion falls back within the inner cone and is also discharged by another spout, from which an elevator delivers it again to the mills. The degree of fineness is regulated by altering the speed of the fan and by adjusting the current of air, by means of valves or dampers at the extremities of the blast-pipes.

Plan of Sifting advocated.—Of the two methods, we think that for a moderate degree of fineness, such as that involving the use of a cloth on the sieve, with not more than 50 meshes to the lineal inch, sifting is the best. With finer cloth than this there is considerable difficulty in keeping the meshes clear, and a separator will be found most suitable. The core or unground material from either sieves or separators should be carried by an elevator to the feed-hoppers and distributed over the crushed clinker, as it is not readily ground alone by mill-stones dressed in the manner previously described. It may be mentioned here that the feed-hoppers should be of sufficient size to contain at least twelve hours' supply of materials, so as to avoid using the crushing machinery when grinding during the night.

French Burr - Stones Best Used for Raw Materials.— Before briefly referring to other methods of grinding, we may mention that for raw materials treated by the dry process, grinding by French burr mill-stones is by far the best method. For this process the materials consist of a mixture of limestone, together with clay or shale, often of widely different composition and of varying degrees of hardness; the hardness of some portions of them, however, rarely, if ever, approaching that of cement clinker. Almost, if not all, the other suggested systems of grinding depend on the separation of the fine and coarse particles, and the regrinding of the latter. The importance of obtaining a perfect mixture of the materials has been referred to in a previous section of this book, and it will be obvious that a system of separating during the grinding is not calculated to attain this end. It has been suggested that the different kinds of material might be ground separately, and afterwards mixed in their proper proportions by suitable mechanical contrivances, but even this plan has many objections.

Raw Materials should be crushed and Ground Together.
—Our own experience, in fact, goes to show that the best
results are obtained by both crushing and grinding the raw
materials together, a very perfect mixture indeed being obtained
by the action of the mill-stones. In comparing different systems
of grinding, it should be remembered that the cost of grinding
the material under consideration by mill-stones is not nearly so
great as that of grinding cement clinker, as the mills take much
less power to drive, and the mill-stones do not require redressing
nearly so often, and they consequently last much longer. Any
saving of cost by another method of grinding would probably be
lost by the expense of the subsequent mixing.

Mill-stone Grinding almost Universal in Thames District.
—The reason why other methods of grinding have not found
much favour in the Thames and Medway districts is probably
due to the fact that there the system of grinding by mill-stones
and the successful design of the necessary plant is well understood, while the cost of keeping the plant in repair is not great
if it be properly designed. The cost, too, of mill-stones has of
late years been much reduced. In many works in other parts
of the country these conditions do not prevail. Much of their
machinery has been designed as if for grinding corn, the result
being a small output, obtained at a great cost, with never ending
trouble and stoppages for repairs. Works of this class have
been a favourite hunting ground for the makers of new types of
grinding machines, which may be broadly divided into two
classes; those grinding by percussive action at high speeds, and
the roller and edge-runner mills. Of the former kind, we consider Askham's Centrifugal Pulveriser to be the best; this,
together with the Globe Mill and the Sturtevant Mill, have
received considerable attention from cement manufacturers, with
the result, however, as we have seen, that they have rarely been
able to hold their own against mill-stones, owing to the great
wear and tear and the difficulty of obtaining a soft floury product.

Askham's Pulveriser.—In the patent centrifugal pulveriser
manufactured by Messrs. Askham Bros. & Wilson, of Sheffield,
the clinker, after having been crushed, is fed into the grinder by
means of an automatic feeding arrangement, which serves, at the
same time, to adjust the quantity dealt with. On entering the
grinder, the material is distributed in an even thickness over
the faces of two toughened steel rings, or grinding paths, by
arms or stirrers, and it is then pulverised by the centrifugal
force of four steel balls or rollers working on the inner face of
the rings. Our illustration shows the complete plant arranged
by this firm and includes the separator, to which we have already

drawn attention when speaking of sifting machinery. It is claimed for this apparatus that, in proportion to the output, the power absorbed is not much more than half that required in grinding by mill-stones or edge-runners, and that great economy of labour results from the fact that the machinery works automatically, and there is no dressing of stones. Another matter

Fig. 10.—Complete Cement Grinding and Sifting Plant of Messrs. Askham Bros. & Wilson.

which must not be lost sight of is the fact that all working parts, which are supplied in duplicate and are interchangeable, can easily be replaced by any ordinary workman. Moreover, it would seem, on the results of some recent tests, that the quality of the flour produced by this mill is identical with that obtained from burr-stones. From some trials made by Mr. A. C. Carey, M.I.C.E., it appears that the output of cement, ground to a fineness of 6 to 8 per cent. residue on a 50 × 50 mesh sieve, was $4\frac{5}{8}$ tons per hour, the I.H.P. being equal to 8·54 H.P. per ton of cement ground per hour. Some of the large firms in the London district have pronounced a very favourable opinion concerning this plant, which seems to us likely to become a formidable rival to the system of grinding by means of mill-stones.

The "Globe" Mill.—This mill has, we believe, been used with a fair measure of success in some cement works on the Medway. The crushed clinker should be raised into a collecting-hopper with a storage capacity of about 25 tons. From this larger hopper it passes into a smaller one below it, and from thence it finds its way into the feeding-tray, leading direct into the mill. The movement of the tray is regulated at will by the attendant, by means of a lever with link motion placed in a handy position in front of the machine. The apparatus for effecting the grinding consists of large discs revolving vertically. These discs are adjustable, and are constructed to lightly grip a ball weighing about 240 lbs., which they cause to revolve on an annular path, crushing the material to the required fineness. The wear on the ball is stated to be very slight, thus in grinding 800 tons of clinker the loss in diameter was about $\frac{7}{8}$ inch. This reduction in size is met by bringing the discs closer together, which can readily be done by means of nuts on the outside of the mill. There is also between the discs a flexible buffer, which gives a certain amount of play, allowing for the rise and fall of the ball.

Fan Employed to Remove Fine Dust.—By the agency of a peculiarly constructed and powerful fan, making 1,100 revolutions per minute, the powdered material is impelled up a shaft in a uniform cloud. No screen of any kind is employed, but the slower the velocity at which the fan is worked, the greater is the degree of fineness of the powder obtained. The guaranteed scale in the case of ordinary working is stated to be 15 per cent. residue on a 100 × 100 mesh sieve, but it is possible to attain even better results than these. The exhaust-shaft conveys the dust into a large receiver, this in turn leading into a second, and then into a third chamber; or one large chamber may be employed, fitted with baffle-plates to grade the powder. These receivers are hopper-shaped at the bottom, so that the cement may be at once discharged and filled into casks or sacks.

Output from Globe Mill.—Concerning the output of this mill, the following particulars are given by the makers :—The company guarantee the production of 25 cwts. per hour of cement with a residue not exceeding 15 per cent. on 10,000 meshes to the square inch, but it is asserted that in certain cases 35 cwts., and even as much as 49 cwts. per hour have been ground. It is stated that the balls will run without change for 600 to 800 tons of cement, and they cost 30s. per cwt. (about 64s. each ball). The time needed to change the ball is one hour. Concerning the quality of the flour, it is stated that a weight of 118 lbs. per bushel is readily obtained from a good sample of clinker, and that a sample weighing 125 lbs. per bushel filled through a hopper has actually been obtained. We regret that we are unable to reconcile this statement with the facts respecting the fineness of the grinding, as they are not in accordance with our own observations on finely ground cement.

Edge-Runner Mill of Dutrulle & Solomons.—We have still to refer to roller-mills, but we will first speak of machines of the edge-runner type. Of these Dutrulle & Solomons' machines, and those of Neate, have now been for some considerable period in use, and the results of their working have from time to time been published. In the former mill the runners, four in number, are vertical and revolve on different paths. The material to be ground is fed upon the upper one, and by the action of scrapers it is carried down to the lowest level, from which it is delivered to elevators. These conduct it to a sieve which separates the fine and coarse portions, the latter passing back again to the mill to be re-ground.

Runners and Beds.—The runners weigh about 6 tons each, and are built up of specially hard iron, the weight being obtained by cement-filling. The beds are similarly constructed of hard iron, in fact the manufacturers assert that these mills will grind 10,000 tons of cement before the wearing parts need renewal. These results are attained by the selection of specially tough iron and by the great thickness ($1\frac{1}{2}$ inches) of the metal. It is claimed, moreover, that the grinding is effected rather by the rubbing together of the masses of clinker than by the contact with, and consequent abrasion of, the metal surfaces.

Practical Tests with this Mill.—It was found in the course of some trials at Messrs. Booth's works near Rochester that 40 H.P. was required to drive this mill for an output of $39\frac{1}{4}$ cwts. of cement in half an hour, or at the rate of about 10 H.P. per ton per hour. During this experiment the amount of material returned to be re-ground with a sieve of 150 meshes to the lineal inch was 40 per cent., with an 80 mesh sieve 24 per cent., and with a 50 mesh sieve $12\frac{1}{2}$ per cent. The mill was driven at 20 revolutions per minute, the engine running at 65. The ground cement had a weight of 117 lbs. per bushel. In all mills of this type our experience shows that the character of the meal produced differs from that yielded by burr-stones, and manufacturers who have this mill in use have, we understand, some difficulty in turning out a cement of high tensile strength.

Comparative **Tests** against **Millstones.**—In a circular issued by the manufacturers, some trials of this mill are compared with the results obtained by mill-stones. The output per hour is here stated to be 4 tons, ground to a fineness of 5 per cent. residue on a 50 × 50 sieve, the net indicated horse-power required being 60. The case of mill-stones is, however, not fairly put. In one instance, the product from a pair of 4 feet 6 inch mill-stones, ground to the same fineness, is given as 15 cwts. per hour, in another as only 12 cwts. This is simply absurd, as we know from our own experience that, with a good sieve, 27 cwts. per hour can be, and is regularly obtained of this fineness from hard burned clinker per pair of 4 feet 6 inch stones. It is, moreover, claimed that cement ground by these mills is of better quality than that produced by mill-stones, a statement which is at least open to question.

Neate's Dynamic Grinder.—In Neate's "Dynamic Grinder" four runners revolve round an inclined raceway. The clinker is in the first place crushed by powerful rolls, from which it is elevated to the storage-bin, passing therefrom to the mills. From the mill by a system of scrapers, which form a special feature, it is delivered to two elevators. These discharge it into two rotating sieves, the wire cloth of which is protected on its inner surface by perforated steel plates. The fine portion of the ground material is carried away by screws, that rejected by the sieves passes back to the mill. The result of a recent extended trial gave a regular output of 5 tons per hour, the net I.H.P. required being 48·9, and the residue on sieves of 50 and 75 mesh being respectively 6·7 and 17·9 per cent. It is not asserted that cement ground by this method is better than mill-stone cement, but only that it can be in no way distinguished from it. Both the mills previously noticed are constructed for dealing with large quantities of material.

Wood's Grinding Mill.—Wood's grinding mill, also of the edge-runner type, is adapted for a moderate output, the quantity dealt with being about that obtainable from a pair of 4 feet 6 inch mill-stones. It has for some years been largely used for grinding phosphates and basic slag, and it has been recently introduced for cement. In this mill two edge-runners, with bevelled edges, revolve vertically round an inclined raceway, the required degree of fineness being attained by an easily adjusted separator, to which the ground material is carried by an elevator.   It is stated by the makers that cement clinker can be ground at the rate of 25 cwts. per hour, so that 90 per cent. will pass the 50 × 50 mesh sieve.   For use with these fine grinding mills a special form of crusher is made.   This is really an improvement on the edge-runner riddle-bottom mill, in use so long on the Medway for grinding gray lime, and more recently employed for reducing the shale for brickmaking by the semi-clay process.   Used as a crusher, 4 tons an hour may be passed through, none of which is larger than ¼ inch, and, if this is treated by the separator, from 8 to 10 cwts. of fine material, 85 per cent. of which will pass the 100 × 100 mesh sieve, may be obtained.

    Ball-mills.—Some Continental cement makers have spoken in high terms of the performance of the ball-mills, and of these the machines of the Grusonwerk, Magdeburg - Buckau, may be taken as a type.   The chief advantage of this system of grinding is the economy of working and the great saving of labour and power, as compared with ordinary mill-stones.   The construction of this machine is simplicity itself.   It consists of a steel drum, on the interior surfaces of which are plates made of the Gruson chilled cast iron, which is extraordinarily hard and tough.   The drum revolves on a steel shaft, and contains within it a number of steel balls, having various diameters, which are partly carried round with the drum as it moves, but which, in consequence of the interior lining being stepped or corrugated, constantly fall and so crush and triturate the material fed into the mill for grinding.

Fig. 11.—Transverse Section of Ball-mill.

The exterior of the drum is surrounded with a double series of sieves, through which the material passes out, when it has attained the requisite degree of fineness.   The ground cement is then led away by means of the discharge-funnel, the coarser particles being automatically returned for further reduction.   The degree of fineness is, of course, regulated by the outer sieve which may be of any required mesh. The material is fed into the mill by a hopper at one side, at the opening of which are spokes shaped in the form of the blades of a propeller, which introduce the substance into the drum, and at the same time prevent the balls from being ejected. We have reproduced this mill both in longitudinal and transverse sections in the accompanying illustrations.

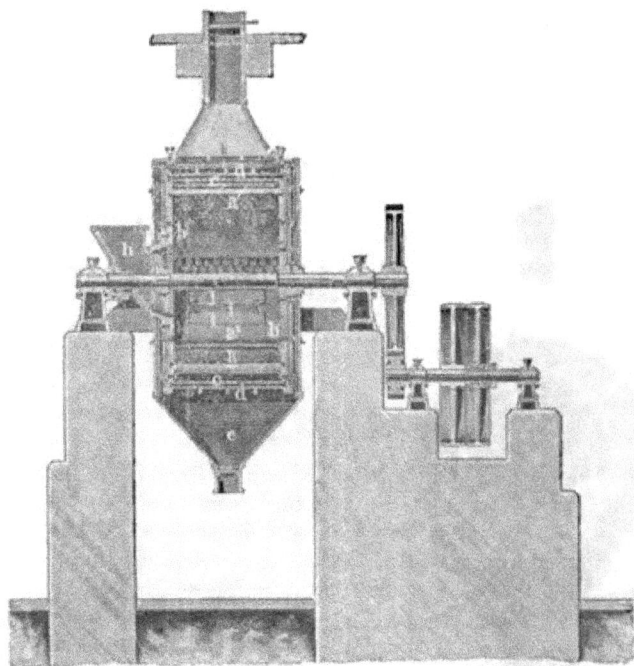

Fig. 12.—Axial Section of Ball-mill.

Mill Grinds and Sifts Simultaneously.—It will be evident from this description that this machine furnishes at once a grinding and a sifting plant, and the whole operation needs so little attention that it is stated that one workman is sufficient for the management of five or six mills. The wear and tear of the parts is said to be very slight, and any defective castings can readily be renewed. It is stated that in a trial with mill-stones at a German factory, the following results were obtained :—

Comparison with Mill-stones.

MILL-STONES.

	Marks.
Fuel for 120 H.P. engine, ...	14·40
Wages, Foreman, ...	5·00
,, Mechanic, ...	5·00
,, Labourer, ...	3·50
,, Two labourers and stonebreaker, ...	5·00
Interest and sinking fund on plant, ...	50·00
Repairs, and wear and tear, ...	30·00
Total, ...	112·90

Equal to £5, 12s. 10d.

BALL-MILL.

	Marks.
Fuel for 90 H.P. engine, ...	10·80
Wages, 1 man in charge of mills, ...	3·50
,, 1 mechanic, ...	5·00
,, 1 workman (for feeding), ...	2·50
Interest and sinking fund, ...	50·00
Repairs, and wear and tear, ...	15·00
Total, ...	86·80

Equal to £4, 6s. 9¼d.

Or, the saving in using the ball-mill, in lieu of mill-stones, amounts to £1, 6s. 0d. per day of ten hours. The cost of grinding a cask of cement with the ball-mill is stated to have been 37 pfennige as against 47·5 pfennige, the cost with common mill-stones. This is a saving of 10·5 pfennige or nearly 1¼d.

Fineness of Resulting Cement.—All of this cement is said to have been ground so fine as to leave not more than 2 per cent. residue on a sieve of 900 meshes per square centimetre, say 2 per cent. on 5,806 meshes per square inch, a standard rarely reached in this country. The turn-out on which these comparisons were made was 40 tons per day of ten hours.

It is pointed out that, if required, the ball-mill may be used to reduce the clinker to the size of peas or shot, and the final grinding can then be done under French burr-stones. When the machine is thus employed a much greater yield may be relied upon, and one or two of the larger size ball-mills will supply the requisite material for several mill-stones.

Novel Process of Grinding.—While on the subject of cement grinding we may briefly notice the process introduced by the British Pneumatic Pulveriser Company, which is both novel and peculiar. The material to be operated upon is fed into a couple of receptacles or hoppers, placed side by side, a short distance apart. Close to the bottom of each of the hoppers there is a small pipe-like opening, the mouths of these openings pointing towards each other. A jet of superheated steam or compressed air is introduced into each opening, and in this way the particles of the material to be pulverised are thrown violently against each other, and the result is a powder of any required degree of fineness, ranging between a coarse sand and an impalpable meal. It is stated that cement clinker may readily be pulverised to a fineness of 200 meshes to the lineal inch by the use of this simple contrivance, and we await further details with considerable interest.

Rollers at one time advocated in Germany.—German cement manufacturers have at times pronounced in favour of the roller-mills, partly because they will turn out more work for a given amount of power, and partly, also, because they require no labour in dressing, and are therefore much more economical than the costly and, to our mind, somewhat antiquated mill-stones.

Flour-Millers in favour of Roller-grinding. — During the last few years flour-millers have increasingly turned their attention to roller-grinding, the flour obtained by this process being, it is stated, superior in quality to, and more bulky than, that produced from the burr-stones; the power required being moreover much reduced. In the case of flour-mills the rollers are made of hard porcelain in the biscuit state, and they are turned to a true surface by diamond cutting tools, and are subsequently polished. In the roller-mills for cement grinding, the rollers are made either of steel or chilled cast iron, preferably the latter, in the form of "Gruson castings," when the wear is much less than with steel.

Experiments by Dr. Tomei.—Some instructive and valuable experiments were carried out in 1880 in Germany, at the Quistorp Works, by Dr. Tomei, which have not, we believe, hitherto been published in this country, but which will serve to put the case of rollers *versus* mill-stones in the best possible light. The material made use of for the tests consisted of the coarse hard particles of clinker, rejected by the sieves. Two pairs of freshly dressed burr-stones, 4 feet 3 inches in diameter, and running at from 110–115 revolutions per minute, yielded on the average of a nine hours' test 8·5 tons of cement per hour, ground so fine that the residue on a sieve with 5,806 meshes to the square inch was only 16 per cent. Twenty-three horse-power was indicated by careful brake experiments, and the temperature of the meal rose from 117° to 192°.

Power needed for Rollers.—Exactly the same amount of power sufficed in the parallel experiment to drive two sets of rollers and one pair of similar stones ; the yield from the rollers was 14·5 tons per hour with a 13 per cent. residue on the sieve of 5,806 meshes per square inch. No observable rise of temperature took place in the grinding and in connection with prolonged use of the rollers, it is stated that after four months working they showed no signs of wear, except that the outer edges were somewhat chipped. The total width of the rollers was 19 inches, but beyond the fact that they were driven at differential speeds, to attain a certain amount of rubbing action, combined with the crushing, which produced a softer meal, no details of the mechanical arrangements are given.

Strength of the Resultant Cement.—The above series of experiments were moreover valuable in that they afforded reliable data respecting the comparative strength of the two samples of cement produced by milling and crushing. In order to test this question thoroughly, samples of the material produced in either way were carefully checked by the sieve and made into briquettes neat and with standard sand. The cement from the stones left a residue on the sieve with 5,806 meshes to the inch of 15·74 per cent., and on a sieve with 32,275 meshes to the inch a residue of 21·15 per cent. ; the total of the coarse particles from both sieves being thus 36·89 per cent. The following table shows the tensile strength of the cement in lbs. per square inch neat and with 3 parts of standard sand :—

	7 days.	28 days.
Neat, ...	428·1	480·7
1 to 3 sand, ...	173·5	240·1

Results were on the Average Uniform.—A sample of exactly the same cement produced by the roller-mill left a residue on the 5,806 mesh sieve of 13·63 per cent., and on the 32,257 mesh sieve of 23·44 per cent., or a total residue of 37·07 per cent. The tensile strength, tested as before, was as follows :—

	7 days.	28 days.
Neat, ...	378·4	492·3
1 to 3 sand, ...	163·6	251·8

It appears from the figures that though the results were a little in favour of the stones at seven days, the twenty-eight days' test gives the advantage to the rollers, and we may take it that, so far as the quality of the cement is concerned, there was practically no difference whatever in the two products.

Grinding by both Processes was equally Fine.—It will

be observed that the grinding was equally fine, but it is pointed out by Dr. Tomei that while the mill-stone ground cement steadily deteriorated from hour to hour, owing to the wear of the dress on the stones, the cement produced by the rollers manifested a slight improvement in quality. The rollers were most accurately turned and polished, and were set so fine and true that a sheet of notepaper, when passed through them, was marked equally on both sides.

Similar Results have not been Obtained in England.— These tests certainly explain the reasons for the confidence at one time expressed by foreign manufacturers in the roller-mills, but we cannot find that any results approaching these in excellence have hitherto been attained in this country. In fact, the following letter received from an eminent English firm at Halling puts a very different light upon this process of grinding :—

Letter from English Users.—We may say that some years ago we erected, at great cost, a complete and most elaborate system of (German) roller-plant on most approved principles. It was a complete failure. It certainly ground the cement to the required fineness on the 50 and 80 mesh tests, but the difference in the comparative strength of the same clinker, prepared by stones alone in the ordinary way, and that when done by the roller-plant was most apparent. Fully one-third less strength neat, and more still when tried with the standard sand test. The setting was always slow, even with an otherwise quick-setting sample. No difference could be ascertained in the appearance or shape of particles under powerful microscopes.

After litigation, the German firm of engineers replaced the rolls by mill-stones, and this gear was only recently taken out and replaced by massive English-made stone gear. The real cause of the failure was the want of flour or impalpable powder in the same proportion as that produced by friction-ground stuff through stones.

Cause of Failure.—It was very fine stuff as a whole, but this flour was wanting, and 80 per cent. or more was really only very fine residue, which acted like sand, and so retarded the setting. Sifting on a 30,000 mesh sieve or through silk sieves told the tale, and these fine sieves are really the only ones that should be used in classifying or testing the values of cement-meal.

Better Results obtained from Edge-Runners.—Since our experience of these rolls, we have always thought that this flour could only be produced by frictional grinding, but last year we saw an English edge-runner mill and tested the same, and found that the flour was equal to that ground through the stones. The tests were also equally good, so there are hopes of the clumsy stone method being superseded. These other systems all take less power than stones.

Manufacturers in Germany mix variously Ground Cements.—Some German friends, who spoke so well of the rolls before we put them down, never had found out our cause of failure until they heard of it here, as it had previously been their habit to mix into one conveyer the product from the rolls and stones, and so the fault had not made itself apparent. We believe that their idea now is the same as ours, and they are agreed that roller-plant without stones will not do, although they can be worked in conjunction with each other. The system was so elaborate, complicated, and extensive, that we preferred to simplify the process, and to do away entirely with the rolls, and not even to have a "joint plant," and we learn that in certain factories abroad, where the rollers were formerly in use, they have reverted to burr-stones.

The Method of using Edge-Runners.—In some factories attempts have been made, as we have seen, to crush and grind the clinker by means of edge-runners of the type of the well-known mortar-mill, and special improvements have been made in this type of mill for the cement trade. But in consequence of the endeavour to make this apparatus do too much, the results have never been very satisfactory. The edge-runner really works much in the same way as the roller-mill, but here the flat pan takes the place of the one roller, which is infinite in diameter, and the runner represents the other. If the clinker, before being placed in the edge-runner, was all broken to a uniform size, we might attain a very fair measure of success, but owing to unequal loading with lump clinker the proportion of really fine meal obtained from the pan is relatively small.

Present Use of Grinding Plant is Unscientific.—The fact is that in using our modern grinding plant it is requisite that each appliance should do its proper share of the work, and do that only. Thus we have stone-breakers and crackers of various kinds, whose office it is to reduce the big lumps of rough clinker to a uniform gauge, say to the size of a walnut. Then the edge-runner, either with a perforated pan or with the runners set at an angle, may be relied upon to prepare a coarse powder for the mill-stones or roller-mills. Even from the pan of the edge-runner a very considerable proportion of fine dust may be obtained by the use of the sieve, and the cement maker must never be afraid of the sieve, which has been far too much neglected in the past, especially in this country. We have been spoiled by the character

of the tests usually imposed, which fail to emphasize the importance of fine grinding, and many cements of German make, when
properly tested with sand, leave our best English Portlands far
behind. It is usual in some cement factories to employ a
graduated series of rollers, and to do all the work of reduction
in this way, and there can be no possible doubt but that roller-
mills do their work very economically.

CHAPTER X.

THE COMPOSITION OF MORTAR AND CONCRETE.

Definition of Mortar. — Mortar, as the term is commonly understood in this country, is applied to the mixture of lime or cement with water to form a plastic matrix for embedding or uniting together bricks or masonry. Such mortar invariably contains a certain proportion of sand. It was an opinion very widely entertained by the earlier writers on building construction that this sand had a beneficial effect on the mortar to which it was added, though we now know that there is no foundation for this belief.

Action of Sand in Mortar.—There are perhaps certain cases in which sand of a peculiar description and used in small quantities may tend to improve a sample of lime or cement, but such cases are quite the exception, for sand can only be regarded in the light of recent experiments as a diluent or as tending to diminish the strength of any given specimen of lime or cement, and the only valid reason that can be urged for its use is the advantage gained in point of economy by the employment of a cheaper material. It has been, however, pointed out that, when exposed to compression in mortar joints, the influence of the hard particles of sand, if duly compacted, may be beneficial, owing to the increased power of resisting a crushing strain of hard particles of quartz, as compared with the friable or putty-like mass of a pure lime mortar or even a cement.

Properties of Mortar.—Mortar suitable for ordinary buildings must have the consistency of a creamy paste, in order to work well under the trowel into the joints and interstices of the materials. It should have considerable power of adhesion, so as to retain the substance with which it is being used in position, and it should possess the power of rapidly indurating and of becoming in course of time as hard as the remainder of the structure, which it should bind together into a species of monolith. Moreover, it should be capable of resisting the attacks of the atmosphere, moisture, and the other destructive influences to which buildings are ordinarily exposed.

Influence of Sand.—It will be obvious, therefore, that,

apart from the qualities of the lime or cement, the sand which enters so largely into the composition of mortars of every description has an important function to fulfil, and though it does not increase the strength of the compound, its action upon the mortar should be carefully studied.

Sand may Act in Four Different Ways.—We may consider its probable effects under four heads. *First*, As respects the cohesion of the particles of the cementing material. *Second*, The adhesion of the lime or cement to the surfaces of the sand grains. *Third*, The possibility of causing the weak places, due to the occurrence of several particles of sand in contact without a sufficient supply of cementing material to envelope them and to fill up all the interstices; *and lastly*, the increase of strength the sand may develop, in such cases in which the sand grains are stronger than the cementing material, in consequence of the line of fracture being longer than it would be if no sand particles were present to interrupt the direct passage of the fracture in one plane.

Sand Tends to Weaken Cohesion of Mortar.—Treating of these various points *seriatim*, we know from numerous tests, conducted by skilled observers, with every description of lime and cement that the cohesion of the particles of mortar made from neat lime or cement is nearly always greater than that which is found to exist between the particles of the same cementing material when used with sand. If we regard the matter simply as one of cohesion or adhesion, though wide differences doubtless exist between the behaviour of various qualities of sand, whether the surface of the grains be rough or smooth, or whether the lime or cement be capable or incapable of exerting a chemical action upon them, the general effect of sand, which retains its position in the mass by the force of adhesion only, is to lessen the force of cohesion. The diminution of strength will be greater in the case of fine sand than it will when coarse or large-grained sand is employed, and it will invariably be larger in direct proportion to the amount of sand we use. For in both these cases the surface to be covered with the cementing agent is enlarged. We substitute, in fact, a lesser force—that of adhesion for a greater force—that of cohesion, the more sand we add. To the same extent, also, that we increase the amount of sand, we incur the risk of weak points from a deficiency of cementing material. The only benefit we can trace then to the employment of sand is the interposition of the sand grains into the direct path of fracture, which they tend to lengthen and increase, and in so far as these grains are stronger than the lime or cement we are using they may improve the mortar.

8

Sand does not, therefore, **improve** the Mortar.—On the whole the advantages of the sand do not counterbalance the sources of weakness it introduces into the mortar, and practical experiments confirm the truth of these views. Mortar is, in fact, merely rendered cheaper by the employment of sand, but not in any sense improved by its use.

Opinions of Early Writers on Sand in Mortar.—From the time of Vitruvius onwards all writers on the subject appear to have taken it for granted that a due admixture of sand must tend to improve the mortar. The earliest builders no doubt found that lime used alone, especially in the form of stucco, had a tendency to shrink and crack in drying, and that this failing could be corrected by the addition of a certain proportion of sand. Having once established this fact, we can well understand that it soon became a settled doctrine that the use of sand was a matter of necessity in the production of a good sample of mortar, but it is strange to find how greatly opinions have differed respecting the reasons for the beneficial effects of this addition, and respecting the proper proportion of sand to be employed.

Proportion of Sand Added.—Vitruvius asserts that "when the lime is slaked it is necessary to mix 3 parts of sand, if the sand is fossil (pit sand), with 1 part of lime; and if the sand is river sand to add 2 parts of it to 1 of lime." He goes on to say that these are the best proportions to use for mortar, and he accounts for the hardening of the mixture somewhat ingeniously as follows:—"The weight of the lime after calcination is diminished about one-third by the evaporation of the watery parts; from this it results that the pores being empty they are better fitted than before to receive the admixture of sand, and to unite strongly with the blocks of stone to form solid masonry."

Theories for Use of Sand.—Many authorities ancient and modern follow Vitruvius and adopt his views; some of them endeavour to invent equally fanciful theories to prove the advantages conferred by the sand. Thus Macquer states that "the particles of slaked lime adhere more closely to hard bodies than to each other, on account of the great quantity of water united to them, which prevents their coming so closely into contact as they do with sand or cement (puzzuolana), for these substances by absorbing a portion of the water of the slaked lime facilitate the drying and adhesion." The amount of water absorbed by sand would be infinitesimally small, and this theory is manifestly incorrect, though it has some show of reason in its favour.

Invariable Rule Established to Add Sand.—It gradually

became recognised as an established fact among builders that lime was capable of taking up a prescribed dose of sand (the amount assumed to be the correct proportion varied greatly), and it was not until Vicat and the careful experimenters of the French school, and our own countrymen, Smeaton and Pasley, began to conduct independent tests that the old faith in the efficacy of sand began to be shaken.

Proper Proportion of Sand to be Used.—In the matter of proportions there has been at all times a great difficulty in deciding exactly what was really intended by writers upon this subject. Some authorities made their mixtures by weight, some by volume, some used lump lime, and others used slaked lime, so that no two results are strictly comparative. The views of Smeaton on the employment of sand are sound and good. He states that: "The use of sand in mortar, so far as I have been able to observe, is two-fold—1st, to render the composition harder; and 2nd, to increase it in quantity by a material that in most situations is of far less expense, bulk for bulk, than lime. As there is no apparent change in the sand by the admixture of the lime, the sand seems only to render the composition harder, by itself being a harder body; for the best sand being small fragments of flint, crystal, quartz, &c., is much harder than any body we know of that can be formed of lime only, which in paste is to be considered as a cement to the harder material, and therefore composes a harder body; for the same reason that if we had nothing naturally but lime as a cement, and should build a wall with flints, crystals, or rough stones cemented therewith, this wall would be harder than if built with lime alone."

Vicat on the Use of Sand.—Vicat, whose experiments on cement were, as we have seen, more careful and elaborate than those of most of his predecessors, cannot be trusted when he deals with the question of sand addition, for he could not in this case shake off the trammels of routine, and he accordingly regarded sand as a necessary part of mortar. He, however, lays down rules for the proportions of sand to be employed with fat and with hydraulic limes, which show a sound appreciation of the subject, and he points out that the quantity of sand must vary in accordance with whether the mortar is to be used in exposed or in protected situations. He asserts that the "intervention of pure sand does not tend or was before believed to augment the cohesion of which every kind of lime indifferently is susceptible, but it is injurious to rich limes, very serviceable to the hydraulic and eminently hydraulic limes, and is neither beneficial nor injurious to the intermediate kinds."

Pasley's Rules for Use of Sand.—Pasley, whose protracted investigations into the nature of cement mixtures and the influence of sand are so well known, did not fully recognise the facts of the case, for he seems to have thought that the strength of mortar was in some way or other related to its plasticity or pleasantness in working. He tells us that he found that Halling lime would not stand so large an addition of sand as the common chalk lime, and he proceeds to state, "every one will acknowledge that the proportion of sand which will make good mortar with chalk lime would entirely ruin cement, which is scarcely capable of bearing one-third of that quantity." Of course he had in view the Roman cement of those days which will take but little sand; but he spoke from the workman's point of view, and regarded a mortar as ruined when it was "too short for use." We now know that Portland cement will make a strong mortar with six volumes of sand, though few workmen care to use it when mixed in these proportions.

Totten's Theory as to Sand Addition.—We do not find any writer on this question who is entirely to be trusted until we come to the experiments of General Totten, who employed a pure fat lime, and who clearly shows that each increase in the proportion of sand involves a falling off in the strength of the mortar; the lime alone furnishing the strongest mortar. His experiments, are set forth in the following table, which we have extracted from the valuable essay of General Scott, published in vol. xi. of the Professional Papers of the Corps of Royal Engineers:—

NATURE OF LIME.	COMPOSITION OF MORTAR. THE LIME MEASURED IN PASTE.						
	0 Sand. 1 Lime.	¼ Sand. 1 Lime.	½ Sand. 1 Lime.	1 Sand. 1 Lime.	2 Sand. 1 Lime.	3 Sand. 1 Lime.	4 Sand. 1 Lime.
Smithfield fat lime,	262·5	245·6	222·5	214·7	170·3	154·6	135·1

The lime used in these tests was carefully ground in a mill and the mortars were moulded into prisms 6 inches long by 2 inches square in section. They were then subjected to a pressure of 600 lbs. for a few minutes, and after 50 days were broken by a force acting midway between the points of support, which were 4 inches apart. Three trials were made of each sample, and the highest breaking weight is in each case recorded. In all the experiments made by Totten he found that the strength of the mortar diminished as the quantity of sand was increased.

Importance of the Size of the Sand Grains.—The size of the sand-grains and the proportion of solids to voids in any given sample of sand is no doubt a subject of much importance in making mortar. It is a matter of common knowledge that if a measure of known size is filled with dry sand, and water is then poured upon it, the sand will settle down as the water is steadily added, and owing to a re-arrangement of the sand-grains the mass will ultimately occupy a greatly reduced volume to that which it does in the dry state. Under certain conditions this shrinkage may amount to 25 per cent. of the volume. In the course of some experiments at Bermuda by Colonel Nelson, R.E., it was found that one part of lime powder when added to three parts of sand scarcely sufficed to fill the voids. In the form of paste it took 1 part of lime paste to fill the voids in 6 parts of sand.

The general experience is that coarse grained angular sand produces a tougher and stronger mortar than a very fine grained sand, though the latter works more sweetly under the trowel. This is probably due to the larger air spaces caused by the cavities or voids of relatively larger area. The voids in sand have been found by careful measurement to amount to about 33 per cent. of the total volume occupied by it. No mortar can be regarded as wholly satisfactory in composition in which the voids in the sand are not filled up by the cementing material, because otherwise certain of the grains might be in contact and have nothing to hold them together.

Scott on Proportion of Sand to be Used.—General Scott, as the result of the careful consideration of numerous specifications, comes to the conclusion that if we are using the feebly hydraulic gray chalk lime, such as is generally employed in the London district, we may safely add $1\frac{1}{2}$ parts or even 2 parts of sand to 1 part of slaked lime by volume. With a more hydraulic lime, such as that obtained from the lias formation, he would use 2 parts of sand to 1 part of slaked lime; and with the pure chalk lime or the fat lime resulting from the calcination of a hard limestone, since the mortar so obtained is at all times very inferior, he states "if we are compelled to use such miserable stuff we shall not be losing much in resistance if we increase the quantity to 3 parts of sand to 1 part of lime."

The Preparation of the Mortar.—All the earlier writers on mortar laid great stress upon the thorough incorporation of the lime and sand, and it would seem by the price-books of the last century that in the country it took quite a day for a labourer to beat together the ingredients necessary to make 1 cubic yard of mortar. At the present time in all important works the mortar is mill-made, and in this way, no doubt, a far better mixture of the sand and lime is obtained than by the old-fashioned system of beating or larrying.

Experiments on the Strength of Mortar.—Many writers have experimented on the strength of the various descriptions of mortar, and it is not our intention to go into this question more in detail in the present chapter, as we shall have to revert to this branch of the subject in our chapter on cement testing.*

Concrete.—We have still to deal with the question of concrete, a term of recent origin, which has been considered by some authorities to be derived from the Latin word *concresco*. This may be taken to mean to grow together, or to consolidate. By "concrete," or the French word "*béton*," is meant a conglomeration of small stones embedded in a matrix of cement or mortar. There are natural examples of concrete in the "pudding stones," where gravel pebbles are united together by oxide of iron, or some similar binding agent, into a hard mass. Building with factitious stone was in use in very ancient times, and some writers have asserted that certain of the Pyramids of Egypt were constructed with artificial blocks of small stones and lime. Many of the earliest buildings in this country, dating back to the time of the Roman occupation, were evidently constructed of a species of coarse rubble-work, wherein fragments of stone and flint are embedded in lime mortar.

Artificial Stone.—As the knowledge of the properties of cements became extended, it soon occurred to ingenious minds that these materials might be employed in the formation of artificial stone, and this idea was first worked out upon a practical scale in France. Towards the close of the last century several inventors made known their processes for moulding a mixture of cement, sand, and gravel. The first experiments in this direction appear to have been due to a Toulouse architect, ·M. Lebrun. He used a specially prepared cement, which he made in a lime kiln; this cement, to which he gave the name of "hydro," was composed of a mixture of lime, clay, and powdered coke or charcoal. The calcined material was ground and mixed with varying proportions of sand into a very stiff paste, with a minimum of water. This paste was at once introduced into the moulds and well rammed with a beetle. The artificial blocks made in this way were sometimes of a considerable size, and arches, with a span of upwards of 30 feet, were constructed under Lebrun's system.

Ranger's Concrete.—In this country the concrete construc-

* In Appendix A (folding table) will be found some experiments by Mr. Kirkaldy on various descriptions of Mortar.

tions of Mr. Ranger about the year 1832, for which he obtained
patents in December, 1832, No. 6341, and again in 1834, No.
6729, attracted considerable attention. The wharf wall of Woolwich Dock Yard was erected by this process, and its subsequent
failure, due rather to insecure foundations than to a defect in
the material, for a time brought concrete work into discredit.*

Vicat in his work on mortars describes a rude system practised
in Piedmont for the production of artificial blocks of stone. Pits
or trenches are excavated in a stiff clay soil to a depth of about
4 feet, and the sides are carefully trimmed to the requisite
dimensions, and thus constitute a species of mould. A mixture
of hydraulic lime and sand is then introduced, and large pebbles
are inserted by hand in such a way as to cause the compound to
be driven into all parts of the mould, the top is levelled off with
a trowel, and then the whole mass is carefully covered over with
about 2 feet of soil and left for two or three years to become
indurated. The lime used for this purpose is obtained at Casal,
and is one rich in magnesia. The blocks formed in this way in
a damp soil obtain a surprising degree of hardness, and may be
dropped one on to the other from a height of 20 feet without
injury.

Concrete in France and England.—In consequence of the
inventions of M. Coignet in France, and of Mr. F. Ransome in this
country, increasing attention has in recent times been paid to the
production of monolithic structures formed of gravel or ballast,
aggregated together into masses by means of lime or cement.
Concrete is really a mortar matrix, serving to bind together a
suitable proportion of pebbles, flints, or broken stones. The
amount of lime or cement to be used in making concrete must thus
be to a certain extent determined by the nature of the materials
to be employed as aggregates. It ought theoretically to consist
of a perfectly made mortar, sufficient in quantity to fill the voids
of the larger materials, for which it is to be used as the incorporat-
ing medium. The proportion of lime or cement may vary from
one-seventh to one-twelfth of the aggregates, but no concrete can
be really sound and good where the materials—the sand and
gravel—are not properly adjusted. We mean by this that the
lime or cement should be sufficient to amply fill the interstices
of the sand, and the lime and sand mortar should properly fill
the voids in the larger masses of stone. It has been found by
careful measurement that while the voids in pebble-gravel will
average about 34 per cent. of the volume it occupies when dry,
the voids in broken stone may average from 40 up to 50 per
cent. of the volume. Taking a mortar composed of 1 part of

* See Pasley's work, p. 21; and again in his Appendix, p. 30.

cement to 3 parts of sand, we might safely add, for the purpose of concrete-making, 3 parts of gravel to 1 of mortar, or 2 parts of broken stone. In some specifications, however, these limits have been dangerously exceeded, and there is sometimes a failure to apportion properly the sand and the larger aggregates. When concrete is merely used as a material for foundations in trenches this is not a matter of very great importance, but where the concrete is to be used moulded into blocks, or in order to form monolithic structures, it is very necessary to study these matters with care.

Experiments of Mr A. F. Bruce on Concrete.—The observations of Mr. A. F. Bruce, who has carried out numerous experiments respecting the strength of Portland cement concrete, and whose investigations are published in vol. cxii. of the *Min. of Proceedings of the Inst. of Civil Engineers*, will show the importance of attending to these details. By careful manipulation of the ingredients and accurate adjustment of the proportions some excellent specimens of concrete known as "*bétons agglomerés*" are produced in France, and have been employed for very numerous purposes, for which stone or terra cotta is used in this country.

Cement Mixing by Machinery.—Machinery has in recent years been introduced into all large works for the preparation both of mortar and concrete, to supersede the former slow and cumbrous method of hand-mixing. In the case of mortar-making many advantages are no doubt obtained by the use of edge-runner mills of sufficient size, especially where lime has to be incorporated with burnt ballast. In the case of cement mortars the employment of the ordinary mill seems to have a tendency to decrease the tensile strength of the mixture, and to retard somewhat the initial set of the cement, but if, for cement mortars, the rollers are taken out of a revolving-pan mill and replaced by a fixed arm with teeth, a very effective mode of mixing is secured, which greatly surpasses hand-gauging.

CHAPTER XI.

CEMENT TESTING.

Early Cement Tests.—The makers of the quick-setting cement of the Roman cement type were fond of displaying the tenacity and tensile strength of that material by the construction of horizontal beams of brickwork, which were built out from a wall a brick at a time. Sometimes a whole day was allowed to elapse between the placing of each brick; at others the fresh bricks were added as soon as the last made joint had become set. For this purpose some five or six minutes would suffice with the best quality of Sheppey cement, and General Pasley observes: "It was stated in my presence at one of the meetings of the Institution of Civil Engineers, by members of that Society, that so many as thirty bricks had been stuck out in one day, and thirty-three bricks in thirty-three days in the same manner." Messrs. Francis had similar beams at their works, and General Pasley tested his artificial cement both by building out the beams with the bricks laid horizontally, and also with the bricks arranged in a vertical position, in which latter case a far greater number could be supported.

Doubtful Value of these Tests.—He says, when speaking of this system of testing: "Considering that this mode of setting out bricks from a wall, though an excellent test of the quality of different sorts of cement, if always done under the like circumstances, might be supposed doubtful, if executed by workmen of unequal skill, or not precisely in the same manner, and in different states of weather or of temperature; and considering also that a pier [beam] of this sort, exhibited at the premises of any cement manufacturer, might not be considered satisfactory by persons who had not been actually present at the time the bricks were placed; since there can be no doubt that, by permanently supporting such piers for several months, a much greater effect would be produced than by the mode I have described of only holding each successive brick a few minutes by a trowel; it therefore appeared desirable when we first prepared for commencing those experimental piers [beams], to make arrangements also for trying the comparative strength of the same from artificial mixtures by tearing them gradually to pieces by a dead weight."

General Pasley's Tests.—He then describes a plan of uniting three (afterwards four) bricks together on the flat by two joints of 9 inches by 4½ inches, having mortises cut in the upper and lower bricks for receiving iron nippers. After allowing from thirty-six to forty-three days for the induration of the cement, the joints were torn asunder by gradually applied weights, laid on by half a cwt. at a time, in an apparatus which was in the nature of a gigantic scale-pan. It was found that when only three bricks were used the weight generally fractured the uppermost bricks by tearing them apart at the mortises, so that in the later experiments four bricks were used, and the mortises were cut in the top and bottom bricks, by which means the solid part above and below the mortises is strengthened by the thickness of one entire brick and a cement joint, leaving the centre joint only exposed to the full action of the breaking apparatus.

Unscientific Nature of the Early Cement Tests.—All of these tests, and even those obtained in later times by joining two bricks together crosswise, so as to obtain a cement or mortar joint 4½ inches by 4½ inches = 20¼ square inches in area (see Fig. 13), were far from scientific, and it was not until recent times that the more accurate and reliable system of using briquettes having a standard area at the neck or point selected for fracture came into use. These notched briquettes appear to have been first used by French engineers. For an early test of Portland cement briquettes of this kind, see the experiments at the 1851 exhibition, p. 34.

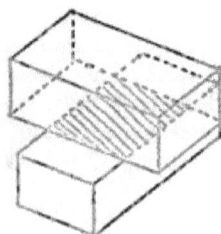

Fig. 13.—Bricks joined crosswise for Cement Testing.

The Value of Portland Cement made known by Testing.—The estimation in which Portland cement is at present held, the reputation it enjoys among engineers, and indeed its high position as a building material are mainly due to the tests it has undergone in the past, and to its behaviour under others of like character and of increasing severity that are constantly being applied to it. It was pointed out many years ago at the Institution of Civil Engineers that, in introducing a cement of this nature, it was very necessary to adopt some general standard of excellence, and the first record we can find of any systematic experiments with Portland cement were those conducted by the Ingénieurs des Ponts et Chaussées, by whom this material was employed on a large scale about 1848 to 1850.

Early French Tests of Portland Cement.—The French tests in use at that period were in the main as follows :—

Specific weight 1,200 kilogrammes per cubic metre, or 103 lbs. per imperial bushel.

Tensile strength of briquettes on the area of 2·25 square inches.

		Neat Cement.			1 Cement to 2 parts Sand.		
		Kilogrammes.		Lbs.	Kilogrammes.		Lbs.
In 2 days,	. .	64	=	140	...		...
,, 5 ,,	. .	128	=	280	64	=	140
,, 30 ,,	. .	240	=	530	128	=	280

Tests by Mr. Druce and Mr. Rendel.—Very shortly after this date Portland cement was used extensively on English harbour works by Mr. Rendel and by Mr. Druce at Holyhead and Dover, and these gentlemen subjected the material to searching tests, but it does not appear that its compliance with any definite standard was at that time insisted upon.

Mr. John Grant's Tests.—The engineers of the Metropolitan Board of Works were the first public officials in this country to institute analogous tests to those adopted in France, and to the late Mr. John Grant, M.Inst.C.E., who was mainly responsible for these tests, and who on several occasions presented to the public the results of the long and laborious investigations carried out by him respecting the strength of Portland cement, we are largely indebted for our present knowledge of the properties of this material, and for the vast improvements which manufacturers have been able to effect in its quality of late years.

Mr. Grant's Essay on Cement in 1865.—In a paper entitled " Experiments on the strength of cement," communicated to the Institution of Civil Engineers in December, 1865, Mr. Grant described the steps taken by him during the construction of the southern portion of the Metropolitan Main Drainage Works, in order to ensure the use of cement of the best quality only. He points out that previous to 1859 Roman cement was with but few exceptions employed for the inverts of the London sewers, the arches being set in blue lias lime, and Portland cement being "scarcely ever tried." It was, in fact, until this time chiefly employed as a stucco, and in ordinary building operations, though it had been, as we have seen, used at Dover and elsewhere mainly for the formation of concrete blocks, and it had already been largely adopted abroad.

Mr. Grant's First Specification Based on Tests.—After some preliminary tests with samples of cement procured for the purpose from various manufacturers, and some corresponding trials of cement obtained from the makers in the ordinary way, the following clause was inserted in the specification for the Southern high-level sewer, dated 1859, being the first contract under the Metropolitan Board of Works on the south side of the River Thames :—

"The whole of the cement to be used in these works, and referred to in this specification, is to be Portland cement, of the very best quality, ground extremely fine, weighing not lest than 110 lbs. to the striked bushel, and capable of maintaining a breaking weight of 400 lbs. on an area $1\frac{1}{2}$ inches by $1\frac{1}{2}$ inches, equal to $2\frac{1}{4}$ square inches, seven days after being made in an iron mould of the form and dimensions shown on drawing (Fig. 14), and immersed six of these days in water."

Briquette-moulding Machine.—In order to prepare the requisite test-briquettes, a moulding machine of simple con-

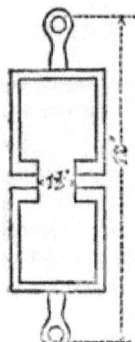

struction was introduced, and bell-metal moulds, having a sectional area of $1\frac{1}{2}$ inches by $1\frac{1}{2}$ inches at the breaking point, were employed. The moulds were originally provided with linings or templates of thin iron, which exactly fitted them and enabled the briquettes to be removed directly they were formed. The moulding apparatus, as also the testing machine, were made by Mr. P. Adie, and these appliances were speedily adopted by all cement manufacturers as well as by most large users of cement, while this system of testing soon became the recognised plan of judging the quality of cement throughout the country.

Fig. 14.
Mr. Grant's
Original Test
Briquette.

Tests made by Ordinary Workmen.—No difficulty was experienced in training an ordinary workman to carry out these tests, and the process was found to answer well in the case of the 70,000 tons of Portland cement used during the construction of the first portion of the outfall works.

Objections to Severity of Test by Manufacturers of Cement.—It seems now somewhat strange to read that in the first instance objection was taken by manufacturers to the standard then proposed of 400 lbs. tensile strength to the $2\frac{1}{4}$ square inches, and it was urged that 300 lbs. would be the highest attainable strength in actual practice. It was soon found, however, that this was not an unduly onerous stipulation for the cement trade, and indeed it shortly afterwards became possible to raise the tensile strength considerably. Thus the clause already given was amended, so that the tensile strength

demanded was 500 lbs., and the contractor was compelled to keep
in store upon the works "a supply of cement equal to at least
fourteen days' requirements," and with each delivery of cement
he was to "send to the clerk of the works a memorandum of the
number of bushels sent in, and the name of the manufacturer."

Modification in Specification about 1865.—Ultimately,
about 1865, the clause was again amended, when it read as
follows:—"The whole of the cement shall be Portland cement
of the very best quality, ground extremely fine, weighing not
less than 112 lbs. to the striked bushel, and capable of maintaining a breaking weight of 250 lbs. per square inch (562·5 lbs. on
2¼ square inches), seven days after being made in a brass mould,
and immersed in water during the interval of seven days. The
contractor shall at all times keep in store upon the works a
supply of cement equal to at least fourteen days' requirements;
and with each delivery of cement shall send to the clerk of
works a memorandum of the number of bushels sent in, and the
name of the manufacturer."

Tensile Strength again Raised before 1870.—Before 1870
the standard strength was again raised, and in the specifications
for the Southwark Park and subsequent works it was stipulated
that the cement should be "capable of maintaining a breaking
weight of 350 lbs. per square inch" (787 lbs. on the breaking
area of 2¼ square inches), tested as before. Thus in a little over
ten years the standard strength had been nearly doubled, and
from the tables given by Mr. Grant it would appear that many
samples were considerably in excess of the specified requirements;
the Burham Company having supplied over 320,000 bushels, showing an average breaking weight of 825·73 lbs. on the 2·25 square
inches; even these figures having been exceeded by certain other
makers who furnished smaller quantities of the material.

Cost of Cement Testing as Carried out under Mr. Grant.
—It is interesting to notice these facts as bearing upon the steady
improvements made in the quality of Portland cement. Mr. Grant
states that the entire cost incurred in testing the cement furnished
during a series of years to the Metropolitan Board, during which
period works to the value of upwards of £1,250,000 were executed,
was about five farthings per ton of cement.

Defects in the Present Mode of Cement Testing.—It is
somewhat unfortunate that the tests first adopted in this country,
and which have thus furnished the precedent for the majority of
engineering specifications, from the fact of their being carried out
with neat cement, fail to accentuate some of the most important
properties of this material, among others the power of binding
together various proportions of sand and other aggregates. The

tensile strength, moreover, **furnishes us with** a knowledge of the
behaviour of the cement under conditions which seldom or never
come into play in actual work, and **the stipulations respecting
weight and fineness** of grinding are **somewhat loosely worded, in**
that they fail **to** indicate with sufficient **precision the manner in**
which the weight **is to be** taken **and the exact degree of fineness
to be attained in the milling, which latter can only be shown by
careful sieve tests.**

Weight Test liable to Considerable Errors.—The mode of
filling the bushel measure, which is one unnecessarily large **for
testing** purposes, is a matter of great importance. This subject
was brought under the attention of the Institution of Civil Engi-
neers during the discussion on Mr. Grant's paper in 1865, by Sir
F. J. Bramwell. **He stated that "he** knew by experience that
when dealing with **granulated matter, it** was almost impossible,
even with the **greatest care, to get** uniformity of conditions in
the respective trials, that was to say, to get the same **amount of
granulated material into a particular measure.** In corroboration
of this he would state the results of what he had tried lately. A
quantity of cement was poured into a bushel measure by a man
accustomed to the work, and struck off level; it then weighed
107 lbs., exclusive of the weight of the measure. **Then another
portion of the same cement was poured slowly out of the sack
down an inclined board into the bushel measure and it then
weighed only 97 lbs.; he then had it shaken down in the measure
and the weight then got up to 132 lbs."**

**Sir F. Bramwell proposed a Specific Gravity Test.—Sir
F. Bramwell pointed out that "when it was found** that with the
same measure of capacity and the same material there could be a
variation of from 97 lbs. to 132 lbs., he could not help thinking,
if a test of gravity could be obtained which was not liable **to
these variations it would be a very desirable** thing." He
further stated that the weight test "should be combined with
that of sifting," an axiom that subsequent enquirers have
amply confirmed. He said that if this was not done, " the re-
quirement as the weight was liable to act as a premium
for coarse grinding." In confirmation of this assertion **he
adduced the following figures:—"He had sifted** through a sieve
of 900 holes to the square inch, a certain portion of the cement
which weighed, as before stated, when very carefully put into
the measure 97 lbs., and when put in in the ordinary way
107 lbs. to the bushel; **those** portions **which** would not go
through the sieve were then poured into the bushel measure
with the same care as that **which** made the unsifted weight, as
previously stated, only 97 lbs., with this care the coarse weighed

as much as 101 lbs., and when shaken down hard into the measure 144 lbs., whereas the unsifted cement had weighed, as before stated, only 132 lbs. He then took by themselves the fine particles which had passed through the sieve and they then weighed only 98 lbs. (93 ?), as against 97 lbs., and when shaken down 130 lbs., as against 132 lbs. for the mixed, and 144 lbs. for the coarse."

From these experiments he was led to believe that with the weight test, in its present form, "there might and probably would be badly ground cement present in the mass."

Specific **Gravity ascertained by Sir F. Abel.**—From some further tests on this same cement made by Sir F. Abel at Woolwich, it was ascertained that the specific gravity was 3·11, equal to a weight of 249 lbs. per bushel, which would seem to show that if the cement could be obtained in a really solid form it would weigh 249 lbs. per bushel, the observed difference being solely due to the voids.

Specific Gravity Test not Suited for General Use.—This suggestion respecting a specific gravity test is a matter of considerable importance, and though this plan has received much attention from subsequent experimenters, it has never become popular, as it is rather too refined a test for actual practice, as Sir F. Bramwell himself, indeed, appeared to think.

The specific gravity test, moreover, has been shown to fail to indicate reliably the degree of calcination to which the cement has been subjected, and it certainly furnishes no guide as to the quality of the material. It has been pointed out by German chemists that Portland cement probably acquires its relatively high specific gravity before it is completely burnt; certainly before the stage of incipient fusion is attained, at which time, owing to the formation of a slag-like mass, we might have expected the molecules to be drawn together more closely, leading to an increase in density.

For a convenient method of ascertaining the specific gravity of cements, the plan proposed by Mr. Keates, involving the use of a double-bulbed bottle, has much to recommend it. We have reproduced in Appendix D (p. 212) the description of this apparatus and the mode of employing it from the *Minutes of Proceedings of the Institution of Civil Engineers*, vol. lxxii.

Form of Early Test-Briquettes.—All of the earlier tests under Mr. Grant were conducted, as we have seen, with briquettes having a sectional area at the point of fracture of 2¼ square inches. Mr. Reid in his work on cements* takes the credit

---

* *The Science and Art of the Manufacture of Portland Cement*, . . . by Henry Reid, C.E. London, 1877. 8vo.

for having suggested the pattern first used; this was avowedly copied from French specifications which involved the use of a briquette of the shape shown in Fig. 14, the notch forming the neck to receive the clips being cut out by hand from a solid casting after the lapse of a prescribed period of time. This must have been a somewhat clumsy plan, necessitating considerable waste of time and money, and it was, moreover, very liable, we should think, to lead to erroneous results in the tests, due to injury to the briquette during the cutting process. The French tests, on a sectional area of $2\frac{1}{4}$ square inches, were of course based upon English measurements for the benefit of manufacturers in this country who at that period were alone able to supply the cement.

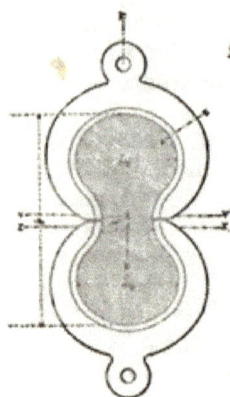

Fig. 15.
Briquette with Sectional Area at Breaking Point of 4 square inches.

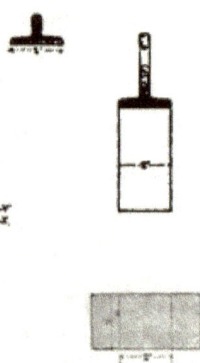

Fig. 16.
Modified Arrangement for Testing Briquettes.

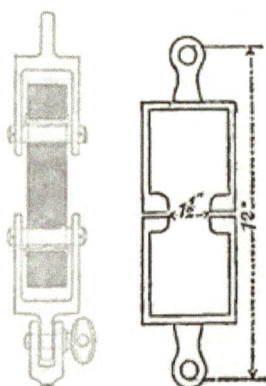

Fig. 17.
Improved Pattern of Briquette.

Proposal to increase the Sectional Area of Test-Briquette.—At the time of reading his first paper on the subject in 1865, Mr. Grant was of the opinion that it would be advisable to modify the shape of the test-briquette and to increase the area at the point of fracture to 2 inches × 2 inches = 4 square inches. The new form of briquette proposed for this purpose being shown in Fig. 15, with shackles made to fit the same. This new shape was, however, speedily abandoned in favour of that shown in Fig. 16. It will be seen that the clip in this case passes through an aperture in the solid ends of the briquette, small castings being inserted in each aperture so as to provide a knife-edge bearing for the shackles to strain against. These frequent changes of form and the many efforts made to obtain reliable results in the early history of cement testing are extremely interesting.

RESULTS OF EXPERIMENTS ON DIFFERENT FORMS OF MOULDS. WEIGHT
OF PORTLAND CEMENT USED, 114 LBS. PER BUSHEL. OCTOBER AND
NOVEMBER, 1866.

No.	Form.	Dimensions at Breaking Point.	AT 7 DAYS.			AT 30 DAYS.		
			Breaking Weight.	Breaking Weight per Square Inch.	Average per Square Inch.	Breaking Weight.	Breaking Weight per Square Inch.	Average per Square Inch.
			Lbs.	Lbs.	Lbs.	Lbs.	Lbs.	Lbs.
1		$1\frac{1}{4}'' \times 1\frac{1}{4}''$	778·1	345·8	...	1010·7	449·2	
2		ditto	1034·7	459·9	...	1151·6	511·8	
3		ditto	1064·3	473·0	...	1145·4	509·0	
4		ditto	918·9	408·4	...	969·0	430·66	
5		ditto	924·0	410·6	...	967·0	429·8	422
6		ditto	860·0	382·2	...	912·1	405·4	
7		$2'' \times 2''$	1199·2	300·0		1354·7	338·7	
8		ditto	1172·6	293·0	297	1315·1	328·8	337
9		ditto	1190·5	297·6		1371·2	342·8	
10		ditto	1286·3	321·6	...	1460·3	365·0	

Influence of the Form of Briquette on the Tensile **Strength.**—In his second paper on the strength of cement, dated April, 1871, Mr. Grant gives a valuable table which we here reproduce, showing the results obtained by the use of the various forms of briquettes employed prior to 1866. It will be seen that ten different patterns were tried, the best results being obtained with Nos. 2 and 3, and it was, therefore, decided to employ briquettes of the former of these shapes for all future experiments. Many of these patterns were tried in order to discover the best mode of avoiding any distortion or departure from the line of strain. The plan originally adopted with this object was that shown in Fig. 15, p. 128, subsequently altered as seen in Fig. 16. It was found, however, that this modified arrangement frequently broke at the holes at each end, instead of at the neck, and it was in the attempt to avoid accidents of this kind that knife edges were inserted in the eye and pin at each end. The fiddle-shaped mould, No. 10 in table, did not break uniformly at the neck, the slightest distortion causing it to break obliquely in a diagonal line. Distortions and irregular strains were often caused by the slipping of the clips, or by the tendency of certain of the clips to open when the strain was applied, and in order to obviate this the clips were strengthened. The form first used with the sharp angles, shown in Fig. 14, p. 124, often broke very irregularly owing to imperfections in the bearing surface of the clips, and this evil was reduced to a minimum by rounding off the edges, as seen in the improved pattern, Fig. 17.

New Shape of Briquette not a Success.—As will be evident from the table, none of the improved shapes tested about this time gave results that were fairly comparative with those obtained with the original pattern, and this form was ultimately adopted for all subsequent tests, comprehended in the series covered by Mr Grant's earlier papers, but in reverting to the use of the original clips, knife edges were inserted in the eye and pin at each end.

Area of Briquette at Point of Fracture.—Following the precedent of the German experimenters, the size of the test-briquette in common use in this country has in recent years been reduced, and it is now customary to employ briquettes having a sectional area of only 1 square inch at the point of fracture. Mr. Grant in his paper on Portland cement, read before the Inst. of Civil Engineers in May, 1880, gave some excellent instructions for the testing of cement, which we herewith reproduce, because his long experience entitles him to speak authoritatively on this matter. He had by this time been converted to the value of the sand test and the use of small briquettes. The form employed by him in his later experiments is shown in Fig. 18.

Instructions for Preparation of Test-Briquettes.—The following were Mr. Grant's instructions for making the tests :— Take enough cement to make as many briquettes of either neat cement or of cement with sand, as may be required. For neat, about 3·20 lbs. of cement will be required to make ten briquettes of the smallest size of 1 inch sectional area, and 16 lbs. for ten of the larger size of 2·25 square inches area. About 1 lb. of cement and 3 lbs. of sand will make ten of the 1-inch briquettes. 500 grammes of cement, 1,500 grammes of sand, and 200 grammes of water will make ten of the German normal briquettes and allow for waste. For testing the cement in different sacks, casks, or bins, a sample may be taken from each and numbered. For experiments, or to get averages, the samples may be taken in equal quantities from each and mixed. The cement, unless it is going to be used at once, ought to be spread in a thin layer on a slate or piece of wood, and exposed to cool dry air as long as may be necessary. With newly-ground and quick-setting cements it is important to ascertain that they are fit for immediate use. Two cakes of neat cement 2 or 3 inches in diameter, about $\frac{1}{2}$ inch thick with thin edges, should be made, and the time noted in minutes that they take to set sufficiently to resist an impression of the finger-nail. One of these cakes, when hard enough, to be put into water and examined from day to day to see if it shows any tendency

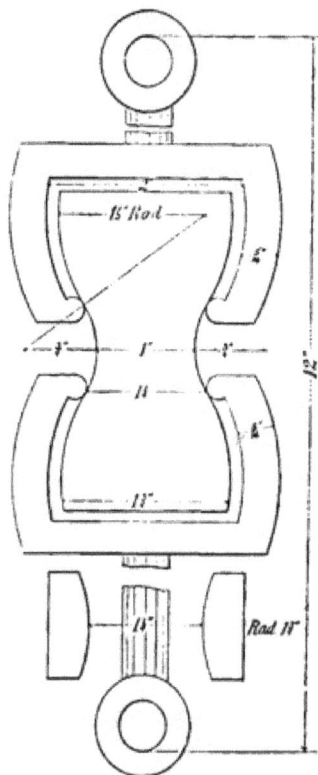

Fig. 18.
Briquette having Sectional Area of 1 square inch.

to "fly," by cracks of the slightest kind, beginning, and being widest at the edges. With slow-setting cement, however, cracks on the surface, beginning at the centre, are merely the result of the surface drying too rapidly from exposure to a draught or to external heat. The other cake to be kept in air

132   CALCAREOUS CEMENTS.

and its colour observed. A small quantity, which may be 1,
5, or 10 lbs., or 10,000 grains = 1⅔ lbs., to be sifted through the
three sieves of 2,580, 5,806, and 32,257 meshes to the square
inch, and the proportion by weight that will not pass through
each sieve tested. The cement to be weighed, using a filler of
one of the forms, Figs. 19 or 20, either, 1st, in an imperial
bushel measure, or, 2nd, in a box = 1/16 of a bushel, which may be
a cube of 6 inches by 6 inches by 6·16, or a cylinder of 6½ inches
diameter and 6·684 inches in depth. The weight to be taken in
lbs. and decimal parts. Ten briquettes to be made with three
times the weight of the cement, of sand which has been washed,

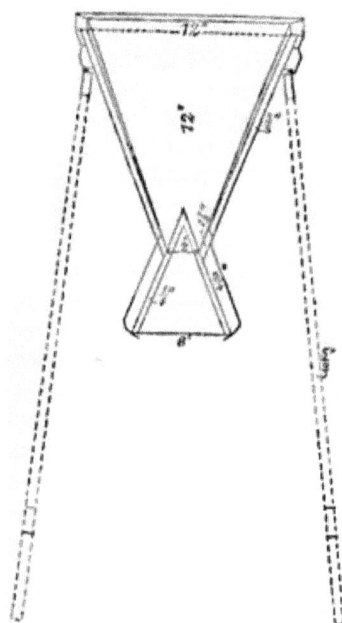
BUSHEL.

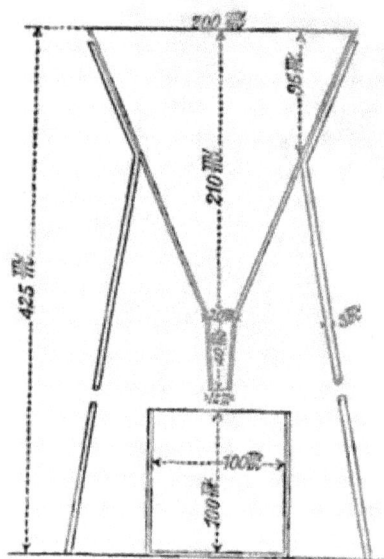
LITRE.

Fig. 19.—Filler for Bushel Measure.   Fig. 20.—Filler for Litre Measure.

dried, and passed through a sieve of 20 and caught on a sieve of
30 to the lineal inch. Twenty more briquettes may be made of
neat cement, to be tested as to strength after seven and twenty-
eight days. The proportion of water (as of sand) to be to the
cement by weight, viz., for neat cement from 20 to 25 per cent.
(or more) according to the fineness, the age or other conditions
of the cement and the temperature of the air. With 3 parts of
sand to 1 part of cement, 10 per cent. of the weight of the united
cement and sand will, as a rule, serve. Sufficient water must

be used to make a stiff paste, but no more. When cement is new or hot, lightly burnt, finely ground, or made up in hot weather, it takes more water, but a very little more than is sufficient to make it into a stiff paste will sensibly diminish the strength. With 3 parts of sand to 1 part of cement the quantity of water required hardly varies from 10 per cent. A number of pieces of wet blotting-paper, a little larger than the mould, one for each mould, may be laid on the slate or marble bench, and the moulds put upon them. The moulds are filled with a small trowel, and the briquettes made with the spatula, see Fig. 22, the mortar being beaten till all the air has been driven out and the mortar has become elastic. The surplus is cut off level, and the surface left smooth. Dr. Michaëlis, of Berlin, recommends another system of making briquettes of neat cement, which, however, is not adapted for very quick-setting cements, nor for briquettes made with a mixture of sand. It is known as the gypsum-plate process. The cement is mixed with from 30 to 35 per cent. of water, and poured into moulds resting on sheets of wet blotting-paper laid upon plates of plaster of Paris. The moulds are tapped or shaken. About 50 per cent. of the water is quickly absorbed, and, if necessary, more of the cement is added. The surface having been smoothed with a trowel or knife, the briquettes are dexterously dropped out of the mould. This process is a very quick one, and Dr. Michaëlis claims for it that it leaves only the amount of water which is required by the cement for setting properly, and that greater uniformity is attained than by any other process. The briquettes can thus be made denser, taking more cement, and breaking frequently under a strain about 50 per cent. higher, or even more. The absorbent slab, he compares to the brick or stone with which mortar is in practice used. But to make strong work with cement it is necessary to soak with water bricks or stone before using them. No doubt, by long practice any one may attain to great uniformity, by following this or any other process ; but we have not succeeded in getting greater uniformity by this than we have by the process first described. When the moulds have been filled the briquettes are numbered, laid aside, and covered with a damp cloth till they have set sufficiently to be taken out of the moulds. The briquettes are then put on sheets of glass, or on slates, and laid in a flat box, having a cover lined with several layers of linen, woollen, or cotton cloth, kept damp. In this box they are kept until they have hardened sufficiently to be put into water. This will vary from one to two hours to a day or more ; but for uniformity, unless in cases of specially slow-setting cement, briquettes of neat cement may be kept for twenty-four hours, and those with sand for forty-eight hours, before being transferred from this box to the shallow tanks in which they are to remain till the moment of testing. The numbers on the neat cement briquettes may be made with a sharp point or with a strong pencil. For the sand and cement briquettes, numbers previously written or stamped on small pieces of zinc will be found more convenient. The water and the testing room should be kept at a temperature as nearly uniform as possible, say, from 60° to 70° Fahrenheit; but if the box and tanks in which the briquettes are kept be covered, moderate changes of temperature will not materially affect the results.

It may be useful, especially at certain seasons, or in countries in which the temperature varies very much, to note the readings of the maximum and minimum thermometers. Seven, twenty-eight, or more days after the briquettes have been made, their tensile strength can be ascertained. In doing this as much uniformity as possible is to be observed in applying the weight slowly and gradually, avoiding all sudden jerks. The result for each briquette being recorded, the average is ascertained.

The test with neat cement at the end of seven days is of little more use than to show approximately, by comparison with the tests made at a later date, whether the cement has increased in strength. The later tests are of most value, and especially those made with sand.

Appliances for Testing.—The apparatus and appliances which it is necessary or desirable to have for making tests and experiments are: a number of boxes about 4 inches deep, lined with lead or zinc, having loose covers and a tap for drawing off the water; a similar box, which need not be lined, but the lid of which should be covered inside with several layers of coarse linen, woollen, or cotton cloth, to be kept always damp, and made to fit closely over the box; a number of pieces of plate-glass or slates about 12 inches by 6 inches, or 15 inches by 8 inches; an imperial bushel measure; a measure of one-tenth of a bushel, either 6 inches square by 6·16 inches deep; or, if cylindrical, 6·5 inches in diameter by 6·684 inches deep; scales for weighing a bushel or smaller quantities of cement, sand, or water; weights of various denominations, including a pound and decimal parts of a pound; 10,000 grains and decimal parts; funnels or fillers for the bushel and smaller measures (see Figs. 19 and 20); brass wire sieves of 400, 900, 2,580, 5,806, and 32,257 per square inch (the first two are used for sand, the last three for cement, and correspond with the sieves used in Germany of 400, 900, and 5,000 per square centimetre); ten or more gun-

metal moulds for making briquettes of 1 square inch sectional
area (Fig. 21), and, if required, similar moulds for briquettes
$1\frac{1}{2}$ inches by $1\frac{1}{2}$ inches = 2·25 square inches; machine for ascertaining tensile strain (see Fig. 24, p. 142); a spatula (Fig. 22),
with thin edges to the blade, weighing $7\frac{1}{4}$ ozs., to be used for
beating the cement and sand in the moulds; straight-edges for
striking off the measures; a straight-bladed knife for cutting off
surplus cement in making briquettes; one or two small trowels;
two strong basins or mortars; maximum and minimum thermometers; blotting paper; a graduated glass for water; sand,
washed, dried, and sifted through a 20, and caught on a 30

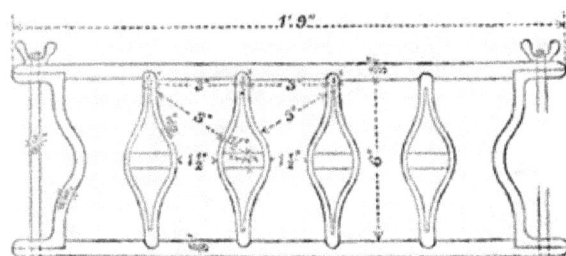

Fig. 21.—Frame for Moulding Five 1-inch
Briquettes.

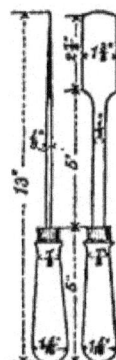

Fig. 22.—Spatula
for Cement Testing.

meshes to the inch sieve = 400 and 900 per square inch; forms
for tabulating and registering results.

Professor Unwin on Cement Testing.—Professor Unwin,
in the course of a paper contributed to the *Journal of the Society
of Chemical Industry* in April, 1886, on this subject, points out that
" testing began with comparatively crude tests of the tenacity of
briquettes or test-pieces of neat cement, which at first were made
of an excessively bad form, and were seldom tested after more than
seven days' hardening." He proceeds to show that the defects
of this plan soon became evident, and caused great differences
in the apparent strength of the same quality of cement, in
accordance with the amount of water used in gauging, the
pressure employed in moulding, and so on.

**Tensile Strength of Briquette depends to some Extent
on the Form.**—Many different shapes were as we have seen
employed, and at one time engineers vied with one another in
the invention of new patterns for the briquettes. As Professor
Unwin says, " the influence of the form of the briquette on the

strength is considerable," different briquettes of the same cement moulded in different forms gave results at seven days, ranging from 280 lbs. per square inch to 460 lbs. on the same area. This question was, as already stated, carefully investigated by Mr. Grant. Even now, when engineers are fully informed of the importance of the shape of the briquette and of the wide range in values arising from trifling variations in the manipulation of the material during the process of moulding the blocks, we are, at any rate in this country, still very far from the possession of any such generally accepted system of conducting the testing process, as shall ensure uniformity in the results. Indeed so many discrepancies in the procedure are to be found that it is exceedingly difficult to get absolutely trustworthy comparative tests. Professor Unwin refers to the care and skill with which cements are examined in the German State laboratories, and it must be remembered that in many parts of the Continent cement producers and cement users have combined to enforce the adoption of a uniform system of tests.

Present Method of Testing Cements.—Cements are now generally tested, either alone, or mixed with certain proportions of sand, the latter tests being by far the most scientific and searching. We must remember, moreover, that cement is tested from two entirely different motives :—

1. To ascertain whether it satisfies certain contract conditions and complies with definite standards laid down in the specification. For these tests the element of time is an important consideration.

2. For scientific purposes, to determine its properties, its value as a constructive material, and the best methods of employing it.

Engineering Tests of Cement.—Under these latter tests the engineer has, as far as possible, to reproduce the exact conditions and surroundings to which the material will be exposed in practice, and his investigations must necessarily be continued for more or less considerable intervals of time. He must assure himself by protracted tests that the cement is a binding agent that may be relied upon for all reasonable periods. The classes of tests which satisfy the first of these requirements are from the very nature of them imperfect and unsatisfactory, and incalculable injury has been inflicted upon cement users in the past, because every effort was made to produce materials capable of giving the best results under these commercial tests, and over-limed coarsely ground cements were preferred to those which, under a more scientific system of testing, would have established their undoubted superiority. We fully endorse

Professor Unwin's declaration that "the great progress of the cement manufacture in Germany, and the production there of cements stronger in test and of more value in the market than English cements is to be traced directly to the adoption of more rational tests."

Fallacies in the Method of Testing.—A cement employed constructively is only in a very few exceptional cases used in tension, but all our English tests would seek to establish its quality by its power of resisting tensile strain. We never, except as a stucco, use Portland cement neat, but it is by the examination of a briquette of neat cement that we too often seek to ascertain its quality. We invariably allow an interval of weeks or months to elapse before we submit our brickwork or masonry to severe stress, but our judgment is in most cases pronounced upon a sample of cement by its behaviour after a seven or at most a twenty-eight days' test. Our walls and buildings have to stand the strain of exposure to sunshine and shower, to heat and cold, and to every change in the weather, but our cement-tests are carefully kept at one even temperature in a "water-bath."

Conditions under which Cement is Actually Employed.—As Professor Unwin remarks, "obviously the ordinary tests are empirical, and may be misleading. Almost every circumstance which markedly affects the strength of a small briquette, tested neat and soon after setting, produces a less effect on the strength of a mass, mixed with other materials, and tested after a longer interval of time."

Qualities needful in a Cement.—He proceeds to show that "the constructional value of a building cement depends on two quite distinct elements—on its power of setting into a rigid form soon after it is gauged, and on its power of attaining in course of time a considerable strength." We may take it for granted that no manufacturer of Portland cement at the present time would think of sending out a material incapable of satisfying the former of these requirements, though we must admit that hot over-limed cements are still occasionally met with which will set well enough for a day or two, but eventually go abroad even in the short space of the seven days' test. Such cases are, however, rare.

The Induration of Cement takes place according to a Uniform Law.—When ample time can be devoted to the study of the quality of any given sample of cement, and when we can note its behaviour through a long series of months or years, we shall find, as Prof. Unwin has remarked, that "beyond the first week, and up to the period at which the full strength of the cement is reached, the rate of hardening follows approximately a very simple law." The factors may be arranged in the form of a readily intelligible formula, the constants of which clearly indicate the character of the cement.

"Let x be the number of weeks during which a test piece has been hardening. Then the strength y in lbs. per square inch at that age is given very approximately by the equation—

$$y = a + b(x - 1)^n$$

where a is the strength at seven days, and b and n are constants depending on the rate of gain of strength with age.

For all tension tests of Portland, n may be taken as the cube root. For ordinary limes it has a larger value. For compression tests of comparatively large blocks of Portland, n is $\frac{2}{3}$; consequently, two tests of a cement at different ages determine the remaining constants of the formula."

The Professor states that the curve given by this formula reconciles the anomalies which must always occur in tensile tests, and produces an average of the somewhat discordant results. He finds by its application that neat cement tests attain their full strength in three or at most four months, while sand tests increase regularly in strength for two or three years. On working out the results from a long series of tests by Mr. Grant, he obtained for neat cement and cement mortar ($1c + 3s$) the following results :—

TESTS OF BRIQUETTES OF NEAT CEMENT AND 1 CEMENT TO 3 SAND AT DIFFERENT AGES OF HARDENING IN POUNDS PER SQUARE INCH.

AGE IN WEEKS.	NEAT CEMENT.		1 CEMENT TO 3 SAND.	
	Observed.	Calculated.	Observed.	Calculated.
1	363	363	157	157
4	415	431	202	214
13	470	471	244	249
26	525	500	285	274
39	512	521	307	292
52	517	543	320	305
104	590	589	351	345
156	585	...	350	372

Specific Use of Prof. Unwin's Formula.—In the foregoing table the value of a in the case of the neat cement is of course 363, and in the case of the sand mixture 157 : the complete formula being—

$$\left.\begin{matrix}\text{Neat}\\\text{Cement,}\end{matrix}\right\} \quad y = 363 + 48 \sqrt[3]{x - 1}.$$

$$\left.\begin{matrix}\text{3 - 1}\\\text{Mortar,}\end{matrix}\right\} \quad y = 157 + 40 \sqrt[3]{x - 1}.$$

In the following graphic diagram for these two substances the actual tests are indicated by dots, the calculated curve is shown by the bold line :—

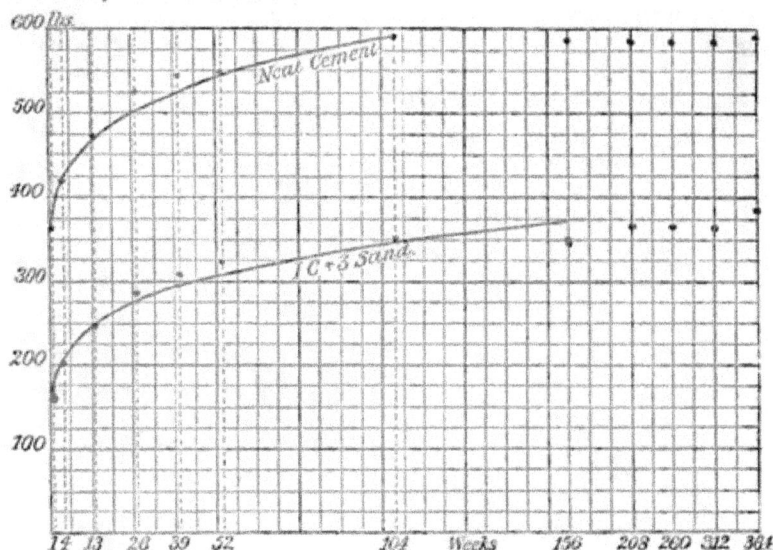

Fig. 23.—Tensile Strength of Cement and Cement and Sand Mortar.

By the application of this same method in the case of coarse and fine grinding, and for various qualities of sand, Prof. Unwin shows how the formula is affected by each different set of conditions. There is no doubt that for careful and comparative tests of cement the application of the graphic method is of great importance, and we commend to all who are interested in cement tests this very valuable and thoughtful essay.

No Definite Limit can be Assigned for the Period of Induration.—It is impossible to assign a fixed limit to the duration of the period within which neat Portland cement may continue to increase in strength. Some cements appear to show a slight increase in this respect up to five or six years, but these cases are exceptional, and in most dense well-made cements the induration has reached its maximum in under two years. After this period the briquettes in certain cases fall back slightly in tensile strength, as if some physical change took place in the structure or arrangement of the particles. Such decrease rarely exceeds 10 to 15 per cent. of the maximum strength.

Value of Long-extended Tests.—It is interesting and valuable to possess records of such protracted tests, and to know the exact composition of the cements with which they were carried out, as it is only by the tabulation of numerous experiments and the comparison of a large number of analyses that we can hope to acquire any accurate knowledge of the somewhat obscure changes which are slowly effected in the course of induration.

Series of Tests of Warwickshire Cement.—We might multiply examples to show this process of "going back" from previous works on cement, but we prefer to adduce a few original and reliable tests to illustrate our meaning. The following experiments exemplify this process in a very remarkable way. They were carried out by Mr. Charles Spackman, F.C.S., with some Warwickshire cement in 1888.

The sample in question was burnt extremely hard, and was of a dark blue colour. It was ground so that it all passed a 50 × 50 mesh sieve.

CHEMICAL ANALYSIS.

Insoluble residue,	.	.	.	.	4·557
Silica, .	.	.	.	.	16·865
Ferric oxide, .	.	.	.	.	4·831
Alumina, .	.	.	.	.	6·360
Manganese oxide,	.	.	.	.	trace.
Lime, .	.	.	.	.	62·507
Magnesia,	.	.	.	.	2·365
Sulphuric acid,	.	.	.	.	1·589
Potash,	.	.	.	.	·976
Soda, .	.	.	.	.	·527

Total,   .   .   .   **100·577**

TESTS OF THE ABOVE CEMENT.

Neat cement, water 16 per cent. The mixture set slowly.

7 days.	28 days.	90 days.	550 days.
Lbs.	Lbs.	Lbs.	Lbs.
515	630	710	695
500	630	760	670
600	620	730	685
660	650	800	660
650	650	780	655
630	660	750	685
620	570	760	660
650	600	790	655
615	620	750	705
690	620	815	675
		Average of 10 tests.	
613	625	764·5	674·5

Cement 1 part, standard (Leighton Buzzard) sand 3 parts by weight, water 10 per cent.

7 days.	28 days.	90 days.	550 days.
Lbs.	Lbs.	Lbs.	Lbs.
185	205	220	310
200	210	215	290
200	220	200	305
200	215	205	335
195	200	215	300
210	215	205	295
200	225	230	300
210	205	200	240
195	225	215	325
200	205	240	325
Averages.			
199·5	212·5	214·5	302·5

Critical Period during Induration.—In the above tests it will be observed that the values for the neat cement indicate a decided falling off between the 90 days' test and that at 550 days, and all observers who have carried out corresponding experiments will have noticed the intervention of such a critical period generally before the end of the second year, especially in very dense cements. No such decrease in strength is apparent in the sand tests, and it is probable that some molecular changes occur during the induration which are not yet understood. This critical period does not always occur after the same interval, and it varies in accordance with the composition of the cement, the temperature at which it has been fired, and the degree of density that it has attained. A well-known writer on Portland cement has attempted to explain away this falling off in strength and to attribute it solely to carelessness in testing. He even says concerning some breaking weights which indicate just such a diminished strength—"It is to be hoped that no more of these records of questionable value will be published, for their misleading and uncertain character has already created too much scepticism on the possibility of cement improving in value."

German System of Cement Testing.—We have referred more than once to the German system of testing, and we give in the case of slag cement a translation of an official report on a series of experiments conducted with that material. In Appendix E (p. 213) will be found a full translation of the Regulations under which such standard tests are conducted. We

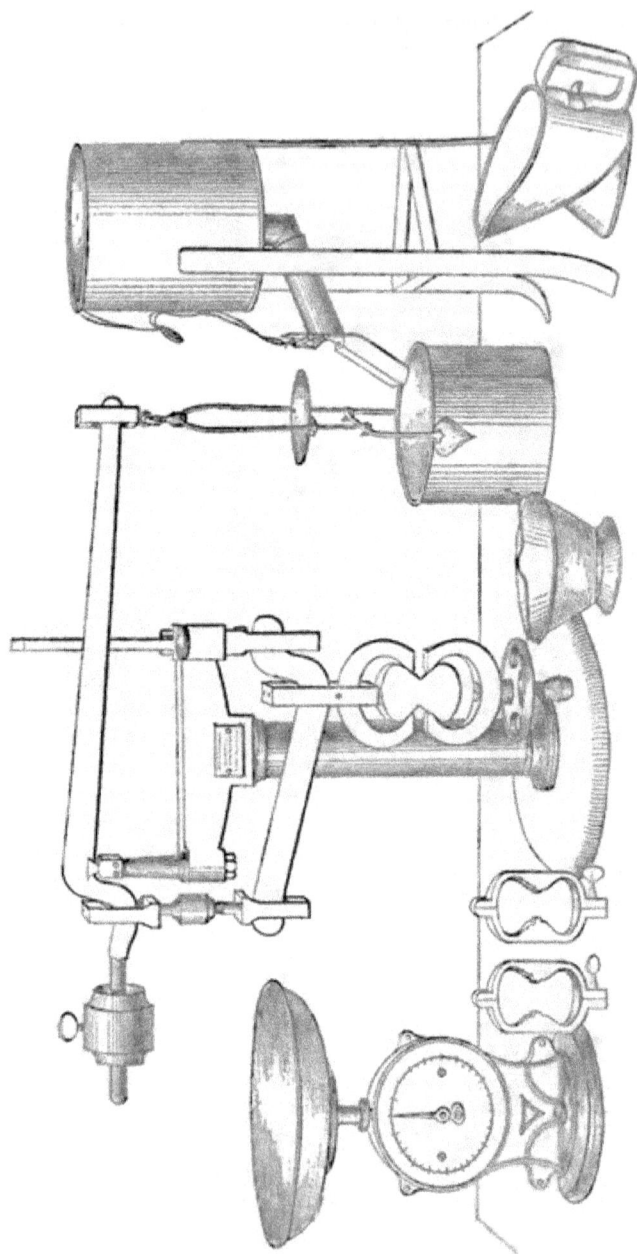

Fig. 24.—Standard Machine employed in Germany and Austria for Testing Cements.

herewith reproduce an illustration of the testing appliances used in Germany and Austria for these official tests.

Arnold's System of Cement Testing.—A method of testing cement, devised by Mr. Arnold, deserves passing mention, as it no doubt eliminates some of the imperfections in our present system which arise from the differences in handling which it is impossible altogether to avoid. We all know that one man will get results some 10 or even 20 per cent. better than another with the same sample of cement. Mr. Arnold's method consists in filling the mould for the test briquette with a proper quantity by weight of the dry neat cement, which is then consolidated under a screw-press, resembling in construction an ordinary office copying-press. The plate on which the moulds rest has shallow sides, thus making it a bath. After removal from the press, each mould having been separately pressed, the bath is filled with water (when full it contains a depth of about $\frac{3}{4}$ of an inch), and the moulds stand therein for twenty minutes before removal. The moulds are made deeper than is required for the section of the briquette. Metal dies which exactly fit the mould are placed upon the top of the cement before it is inserted in the press, so that when the die is properly pressed down, its top is level with the top of the mould and no further compression can take place. Thus a uniformity of pressure is insured in the case of each briquette moulded from the same sample of cement. In order to avoid difference arising from varieties in the bulk of different qualities of cement, it is pointed out that dies of various thicknesses can be used. It is claimed for this system of making briquettes that much greater uniformity is attained than by the plastic method, that the services of an experienced man are not required, that the requisite amount of water and no more or no less is used, and that air bubbles are entirely avoided. The patentee states, moreover, that the evidence of a blowing cement will be noticed in 24 hours after moulding the briquette, and that owing to the absence of disturbance, after admixture with water, any particles of free lime present are invariably detected at the point of fracture.

The Hot-Water Test for Cement.—Increasing attention has of late been given to the method of testing the soundness of cement by means of the hot-water bath, and this system has much to recommend it, especially in cases in which it is important to ascertain speedily the character of any given sample of Portland. The principle which underlies this plan of testing is that moist heat facilitates and accelerates the induration of cement compounds, and, by the employment of boiling water, the setting process is hastened to such an extent that defects, which might take days or even weeks under ordinary circumstances to become manifest, are developed in the course of a few hours. Different observers have suggested somewhat varying methods of carrying out this hot-water test, which is undoubtedly a powerful agent for the detection of latent imperfections in the composition or quality of cements of the Portland type. Mr. Margetts, who has used this mode of testing, states that no less than ten samples of cements obtained from various sources became more or less disintegrated by the boiling process; some of them being reduced to mud and losing all traces of cohesion. It has been asserted that forty-eight hours in the hot bath are equivalent to immersion for seven days in cold water, that is to say that a briquette made of neat cement will show a corresponding amount of tensile strength after the above intervals respectively, or that, in the case of tests made with sand, a seven days' hot test is equivalent to a twenty-eight days' cold test; the percentage of difference of the tensile strengths in a good sample of cement being in both instances but trifling. In applying this test to specimens of neat cement, it is usual to make up small circular pats, about 3 inches in diameter, with just enough water to enable them to be smoothed out with the trowel to a thickness of about $\frac{1}{2}$ an inch in the centre, and a $\frac{1}{4}$ of an inch at the edges. As soon as these pats become fairly firm or set, they are placed in water kept at a uniform temperature of about 180° F., or some prefer 212° F. for seven or eight days, after which period, if the pat still remains firm and unaltered, the cement may be pronounced to be sound and of good quality, and capable of undergoing any of the ordinary engineering tests. If the water in the bath is raised to the boiling point, a much shorter period will suffice. If the tests are made with samples of cement and sand, however, the lower temperature should be employed. If used in conjunction with the ordinary tensile tests, this method should prove of especial importance to the cement manufacturer, who frequently, when making alterations in his materials or mixtures, requires to form a speedy conclusion respecting the resultant cement.

Hot-Water Test by Mr. Margetts.—Mr. Margetts, in the course of a discussion upon this subject at the Institution of Civil Engineers, in November, 1891, urged the value and importance of this test, and gave the following instance of its application to two cements, nearly identical in their composition, one of which stood the boiling process perfectly, while the second disintegrated after the expiration of three hours in water at 212° F. :—

Composition.	Sample No. 1.	Sample No. 2.
Lime, ...	60·67	60·47
Silica, ...	24·86	24·93
Oxide of iron and alumina, . .	12·28	12·42
Magnesia, ...	0·63	0·58
Sulphuric anhydride, . . .	0·36	0·30
Carbonic ,, ...	0·47	0·52
Alkalies, ...	0·73	0·78
Total, ...	100·00	100·00
Specific gravity, . .	3·11	3·118

Tested with the sieve, No. 2 proved to be somewhat more finely ground than No. 1, and the average tensile strength of No. 1 after seven days (six days' immersion in cold water) was 496 lbs. per square inch. The same cement gave, after twenty-four hours in the air and forty-eight hours in the hot-water bath, a tensile strength of 418 lbs. per square inch. No. 2 cement, tested in the ordinary way at seven days, had a tensile strength of 390 lbs. per square inch, but briquettes exposed for twenty-four hours to the air and then immersed in boiling water became disintegrated, as already stated, in three hours.

CHAPTER XII.

THE CHEMICAL ANALYSIS OF PORTLAND CEMENT, LIME, AND THE RAW MATERIALS REQUIRED FOR THE MANUFACTURE OF CEMENTS.

Original System of Mixing the Raw Materials.—In the early days of the Portland cement industry the correct adjustment of the proportion of carbonate of lime to clay was purely an empirical process, and was in fact an application of the system of trial and error. In the first place, mixtures were made in varying proportions, from which samples of cement were prepared, and the proportions giving the best cement were adopted. Any irregularity due to variations in the chemical composition of the materials, or to carelessness on the part of the workmen, was ascertained by sampling the mixture frequently, the samples being burned in a trial-kiln.

Unsatisfactory Character of this Mode of Admixture.— From a close observation of the resulting cement when tested for soundness, colour, &c., the "sampler" was by long experience enabled to judge whether the correct proportions were being maintained. This was a tedious and an unsatisfactory process, for while a sample was being prepared and tested the bulk it represented had passed beyond the reach of alteration. If the slurry was run into backs, any error was corrected by altering the mixture and luting or stirring the contents of the back, from which a sample was occasionally prepared. With the Goreham process no alteration of the previously prepared mixture was possible, it would probably be dry and ready for the kiln before the results of the sample were known. If the contents of any particular drying-flat or chamber proved to be over- or underclayed, the burner could be instructed to burn it lightly or heavily as the case might be, and this was really all that could be done.

Mode of Working by the Percentage of Carbonate of Lime.—When it is remembered that if the materials are in the first place suitable, success depends on the proper proportion of carbonate of lime to clay, and that for the same material this proportion is a constant one, it is evident that, if the percentage of calcium carbonate in a normal sample is known, the correctness of the mixture may be checked by the determination of this substance in the trial samples.

Value of Sample - Kiln Tests.—At the same time the sample-kiln should not be ignored, especially when establishing the manufacture with materials with which the operator has had no previous experience, and indeed the most accomplished chemist when commencing the study of cement-making may learn much from the old-fashioned sampler, his methods of work, and the unconscious instinctive way in which he reasons from the appearance of his samples. It should never be forgotten that Portland cement was the invention of a bricklayer, and had established its reputation long before its manufacture had become a scientific question. We ourselves know of the existence of work executed nearly fifty years ago, with cement obtained from Aspdin's works, which is as perfect now as it was on the day when it was completed.

Scheibler's Calcimeter.—An instrument known as Scheibler's calcimeter, originally devised by Dr. Scheibler for determining the amount of calcium carbonate present in the animal charcoal used for sugar-refining, is frequently employed for making a similar determination in Portland cement mixtures. A description of the apparatus with instructions for its use will be found in several text-books of analytical chemistry,[*] and they need not be repeated here. The principle of its action is this: the carbon dioxide, evolved by the action of hydrochloric acid upon a weighed quantity of the material to be tested, displaces a corresponding volume of water from a graduated tube. To the volume when read off is added a correction to allow for the carbon dioxide retained by the acid used for the decomposition. The total volume is corrected for moisture, temperature, and barometric pressure, and from the corrected volume the weight and percentage of calcium carbonate in the sample operated upon is then calculated.

Method of obviating certain Errors due to this Process.—This method, although it is perhaps "near enough" for many technical purposes, is far from accurate. The carbon dioxide evolved is collected over water, in which it is to some extent soluble. This error may be avoided by the use of Lunge's nitrometer, in which the gas is collected over mercury. The volume retained by the liquid in the evolution-bottle or flask, even when the same quantity of acid of the same specific gravity

---

[*] Fresenius, *Chemical Analysis—Quantitative*, vol. i., p. 344, 7th ed.; Sutton's *Volumetric Analysis*, p. 87, 5th ed.; Clowes & Coleman's *Quantitative Analysis*, p. 356; Crookes' *Select Methods in Chemical Analysis*.

is always used, varies with the temperature, pressure, and the
total volume of carbon dioxide given off. Scheibler directs the
operator to add 3·2 c.c. to the volume read off, when 10 c.c. of
hydrochloric acid of 1·12 specific gravity are used. Warington,
who has experimented with the apparatus,* recognises the fact
that the quantity retained by the liquid is dependent upon the
volume of gas evolved, and he gives 7 per cent. of this total
as the volume to be added.

Alternative Plan of Using the Calcimeter.—If a number
of experiments are continuously carried out upon the same
sample under the same conditions, and if the temperature and
pressure do not vary during the operation, the volume of gas
obtained will in each case be alike, and a better method to adopt
in using instruments of this kind for testing the accuracy of
cement mixtures is the following one :—Two samples of the raw
mixture should be prepared, reduced to a fine powder, dried at
100° C., and preserved in well-closed bottles for use. One of these
samples, say No. 1, should contain the minimum quantity of
calcium carbonate it is advisable to work with, and No. 2 should
contain the maximum, and in each of these mixtures the amount
of calcium carbonate should be accurately determined by analysis.
It is also as well to prepare at the same time a normal sample, the
quantity of calcium carbonate in which would of course be about
midway between those in the other two specimen mixtures. The
same quantity of dilute acid of the same strength should always be
used for the determinations. The volume of gas from the sample,
the correctness of which it is desired to ascertain, having been
read off, determinations should be made with Nos. 1 and 2, and
the results compared. The time occupied is of course greater
than if one determination only was made—viz., of that in the
actual sample to be tested; but, on the other hand, calculations
are unnecessary, and the result obtained is much more reliable.
The volumes ascertained from the samples of known composition,
together with the pressure, or height of the barometer, and the
temperature should be carefully tabulated. Reference to these
tables will frequently obviate the necessity for making more than
the one determination. It is perhaps needless to point out that
the samples to be tested should in all cases be previously dried,
the quantity should be accurately weighed, and the same weight
of substance should always be taken.

·   **Apparatus for the Determination of Carbon Dioxide.**—
An apparatus by which carbon dioxide can be accurately determined volumetrically has been devised by G. Lunge and L.

* *Chemical News*, vol. xxi., p. 253.

Marchlewski,* but it is not adapted for use by those who are not familiar with the methods of gas analysis. Very little knowledge of chemistry or chemical manipulation is necessary for the use of Scheibler's calcimeter or the Lunge nitrometer.

A more Simple Method of Procedure.—A good method of checking the composition of a cement mixture is to determine the loss on ignition of a sample previously dried at 100° C. The loss consists of carbon dioxide, combined water, organic matter, and sulphur, if present, as it sometimes is, in iron pyrites. The following apparatus is requisite:—Balance and weights, desiccator, platinum crucible with cover, small pestle and mortar, retort or tripod stand, iron crucible tongs, and a blast-lamp. This may be either a suitable Bunsen burner for gas or a Berzelius lamp, the blast being obtained by means of a Fletcher blower. Various forms of lamps burning benzoline are in use, the blast being produced by the vapourised benzoline which issues under pressure from the reservoir. One of the best of these is the Hoskin's lamp supplied by Messrs. Griffin. A few porcelain capsules, a water-bath, and a water-oven are also necessary for drying the samples. These last may be kept heated on small paraffin stoves, if gas is not available.

Mode of Testing Slurry.—The slurry of the wet process should in the first place be evaporated to dryness in a capsule on the water-bath, then powdered in a mortar, and further dried for half an hour in the water-oven. The dust of the dry process may of course be at once dried in the water-oven. After drying it, the sample should be immediately transferred to a clean dry corked or stoppered tube, and allowed to cool. A quantity equal to about 1 gramme is transferred to the platinum crucible, which has been previously weighed with its cover. A little practice will soon enable the operator to judge the approximate quantity. The crucible is covered and again weighed with the contents. The weight of crucible and cover on being deducted from this weight give the weight of the quantity taken for the determination. The cover is removed, and the crucible, resting on a pipe-clay triangle, which is supported on the retort or the tripod stand, is heated over the blast lamp, at first gently, and finally to strong redness for about twenty minutes. It is then covered, removed to the desiccator, and when cold it is again weighed. The difference between the first and second weighings of the crucible, with the cover and contents, is the loss on the quantity of material taken, from which the loss per cent. is readily calculated.

*Zeits. f. angew. Chem., 1891, pp. 229-235. Abstracted in Journal of Society of Chemical Industry, vol. x., 1891, p. 658.

$$\frac{100 \times \text{loss on ignition}}{\text{Weight of quantity taken}}$$

Comparison of above Results with those by use of Scheibler's Apparatus.—If the loss per cent. of the samples, prepared as described when referring to Scheibler's calcimeter, is accurately determined, a standard for comparison is at once established.

The platinum crucible should be occasionally re-weighed as it loses weight by continued use. When tarnished, it may be cleaned by rubbing it with the fingers and a little moist sea sand, after which it should always be re-weighed. It is of course necessary that the ignition should be continued until the whole of the carbon dioxide is expelled, and it is advisable, until the operator has acquired confidence in the use of this method, to ignite the contents of the crucible a second time after once weighing, and to again weigh the contents.

COMPLETE ANALYSIS OF CEMENT MIXTURES.

Foregoing Methods are only Approximate.—The methods previously described are only adapted for checking the proportion of carbonate of lime to clay, and are such as may be carried out by those possessing very little chemical knowledge. It is now proposed to deal with methods specially adapted for the analysis of cements and cement-making materials, which are intended for the use of those possessing some knowledge of general analytical chemistry, or for chemical students engaged in cement manufacture.

The Analysis of Hydraulic Limestones, Grey Chalk, and Portland Cement Mixtures.—The sample should be finely powdered, dried for some hours in the steam-oven, and transferred to a well-stoppered tube. When cold the following quantities should be weighed off from the tube :—

1. From 1 to 1·5 gramme for the determination of sand, SiO_2, Fe_2O_3, Al_2O_3, CaO, MgO :—transfer this quantity to a platinum crucible.

2. About 3 grammes for S existing combined with Fe as FeS_2, and for $CaSO_4$. This quantity should be transferred to a beaker. If a qualitative examination has proved the absence of FeS_2, this quantity may be omitted.

3. About 4 grammes for Na_2O and K_2O, to be transferred to a platinum crucible of ample capacity.

4. And about 1 gramme for CO_2, transferred to the evolution flask of the apparatus used for the determination.

Quantity No. 1.—Determination of SiO_2, Fe_2O_3, Al_2O_3, CaO, MgO.—Gently rap crucible once or twice on bench to consolidate contents, which should not be again disturbed, and ignite it, first at a gentle heat and then strongly for from forty to fifty minutes over a blast-lamp. The effect of this treatment is to decompose the silicates by the action of the lime, while the sand practically remains unacted upon. When cold, the contents of the crucible are turned out into a platinum dish and triturated with a little water, until all tendency to set has disappeared, using either a stout glass rod with rounded end, or better, a small agate pestle. Add dilute HCl, and digest for a few minutes at a gentle heat. The crucible may be rinsed with a little dilute HCl and washed into dish. If the contents have not been disturbed they will be found after ignition to have slightly consolidated, and will drop out of the crucible, leaving only a small quantity behind. Evaporate the contents of dish to dryness (a water-bath is safest), and complete the drying by heating the dish in a hot-air oven for two hours at 150° C. When cold, moisten the contents of dish with strong HCl, adding it drop by drop, and using as little as possible, cover with a clock-glass and set aside for fifteen minutes. Add water, stir, break up any lumps if present, filter, washing the precipitate well with hot water. The precipitate which consists of sand and the SiO_2 previously in combination with bases is dried, strongly ignited in platinum crucible, and weighed.

Separation of Sand and Silica previously in Combination with Bases.— The contents of the crucible are carefully transferred to a platinum dish, and heated for some time with a strong solution of Na_2CO_3. The contents of the dish are allowed to settle for a minute or two, and while still hot the clear liquid is decanted over a small filter. The residue is again treated in the same manner with Na_2CO_3 solution, and the operation is repeated until it does not seem to decrease in bulk. The filter is kept well washed with hot water, and is by preference enclosed in a hot-water jacket. The contents of the dish are finally removed to the filter and repeatedly washed with hot water. If the washing proceeds slowly it is probable that the solution while filtering has deposited SiO_2 in the gelatinous state on the filter. In this case a hole should be made in the bottom of the filter, and the contents carefully washed into the platinum dish, the treatment with Na_2CO_3 being once repeated. It is always safest to repeat this treatment, as the SiO_2 has a great tendency to separate as its solution cools. The precipitate is again filtered off, repeatedly washed with hot water, dried, ignited in a platinum crucible, and weighed. Its weight, deducted from that formerly found, gives the weight of the SiO_2 previously combined with bases. The examination of this sand will be referred to further on.

Determination of Iron and Alumina.—To the filtrate from the SiO_2 add a few drops of bromine water and boil. Allow the liquid to cool, add NH_4OH in very slight excess, and again boil the liquid. The precipitate of $Al_2O_3 + Fe_2O_3$ settles badly, but filters rapidly, even without the aid of a suction apparatus, and nothing is gained by allowing it to settle, and washing by decantation. Wash it once or twice on the filter with hot water, setting aside the filtrate and washings, which should be received into a beaker of ample capacity. The precipitate need not be entirely washed out of the beaker. The precipitate is dissolved on the filter with a little hot dilute HCl, the bottom of the paper is perforated, and the filter well washed with hot water. The solution of precipitate and the washings are received in the beaker in which precipitation was previously effected. The contents of the beaker are boiled and allowed to cool. NH_4OH is then added in very slight excess, and the solution with the precipitate is again boiled. The precipitate is filtered off and well washed with hot water, the filtrate and washings being received in the beaker containing the filtrate from the first precipitation. Care should be taken that this precipitate is sufficiently washed, as it retains traces of lime with great obstinacy. A small quantity of the washings should be collected in a test tube, to which a drop of $(NH_4)_2C_2O_4$ is added, the contents of the tube are boiled, and the tube with the contents must then be cooled by immersion in cold water, as traces of calcium oxalate do not separate from a hot solution. If lime is present the contents of the tube are rinsed into the beaker containing the filtrate, and the washing is continued until a sample gives no trace of lime when tested as described.

The contents of the beaker are heated to the boiling point, sufficient $(NH_4)_2C_2O_4$ solution is added to convert all the lime and magnesia present into oxalates. The liquid with the precipitate is again heated to the boiling point, after which it is set aside for a few hours.

Traces of Silica in Filtrate from Main Precipitate.— If the contents of the platinum dish are not thoroughly dried, a considerable quantity of SiO_2 passes into solution, and is precipitated with these oxides. This will be observed after treating the precipitate on filter with dilute HCl and washing into beaker. Gelatinous SiO_2 will appear in the solution, and this is not dissolved on heating. Traces are, as a rule, found even after the contents of the dish have been heated to 150° C., but, if too strongly heated, loss of ferric chloride may occur by volatilisation, or the chloride is converted into ferric oxide which will not be readily dissolved by the treatment with HCl. The trace of SiO_2 should be separated, and its weight, ascertained by one of the following methods, a or b, should be added to that of the principal quantity previously found.

Estimation of Traces of Silica.—(*Method a*) The mixed precipitate of Al_2O_3 and Fe_2O_3 with traces of SiO_2 is dried, strongly ignited in a porcelain crucible, and weighed. The contents of the crucible are transferred to a porcelain dish and heated for some time with moderately concentrated HCl, the lumps being broken up with a glass rod as the digestion proceeds. After continued heating, everything but the SiO_2 will dissolve. The excess of acid is evaporated off, water is added, the contents of the dish are filtered, and the SiO_2 is washed, dried, ignited in a platinum crucible, and weighed. Its weight, on being deducted from that of the mixed precipitate, gives the weight of the Fe_2O_3 and Al_2O_3.

Determination of Iron and Alumina.—The iron may be determined in the filtrate from the traces of SiO_2, by reducing with zinc and titrating with standard bichromate, or gravimetrically as follows:—The filtrate and washings are collected in a porcelain dish, and a sufficient quantity of concentrated solution of NaOH is added to dissolve the whole of the alumina present. Some idea of the quantity required may be obtained by first rendering the contents of the dish alkaline, and continuing to add the reagent gradually, until the pale red colour of the mixed precipitate has changed to a permanent deep red, the contents of the dish being at the same time stirred and heated to boiling. The precipitate is filtered off, washed twice with hot water, and dissolved on the filter with a little hot dilute HCl. The solution, rendered decidedly alkaline with NaOH, is then heated to boiling, the precipitate again filtered off, repeatedly washed with hot water, dried, ignited in a porcelain crucible, and weighed. The alumina in each case is determined by difference.

Estimation of Traces of Silica (*Method b*) and direct Determination of Iron and Alumina.—The precipitate of Fe_2O_3 with Al_2O_3 and a trace of SiO_2 is dissolved on the filter with hot dilute HCl, the bottom of the filter is perforated, and it is well washed with hot water. The solution of the precipitate and the washings are evaporated to dryness, and the trace of SiO_2 is separated as previously described. The Fe_2O_3 and Al_2O_3 may be precipitated together in the filtrate, and the Fe_2O_3 determined as described in (a), or both Fe_2O_3 and

Al_2O_3 may be directly determined. The Fe_2O_3 is separated as in (a), taking care that the NaOH used for the precipitation has been prepared from sodium, and is free from Al_2O_3.

The filtrate and the washings from the Fe_2O_3 are rendered acid with HCl, a crystal of $KClO_3$ is added to destroy any organic matter which might be present and which would interfere with the precipitation of the Al_2O_3, and the solution is concentrated in a porcelain dish. NH_4OH is added drop by drop until the solution is just alkaline, and the contents of the dish are boiled until a piece of moist turmeric paper held in the steam does not become brown. The precipitate is filtered off, repeatedly washed with hot water, dried, ignited strongly over the blow-pipe in a platinum crucible, and weighed.

Separation of Lime and Magnesia.—The supernatant liquid is decanted from the precipitated calcium oxalate into a beaker of ample capacity. The precipitate is dissolved by heating with dilute HCl, the solution is rendered alkaline with NH_4OH and heated to boiling, after the addition of a few drops of $(NH_4)_2C_2O_4$. The beaker is set aside to cool and to allow the precipitate to settle. The clear supernatant liquid is then decanted into the previously decanted liquid. This is then passed through the filter upon which the bulk of the precipitate is to be collected, and any traces of precipitate are washed from the beaker on to the filter. The main precipitate is washed once or twice by decantation, finally transferred to the filter and washed with hot water. The precipitate is dried in a water-oven, separated as far as possible from the paper, which is ignited in a coil of platinum wire, and together with the ash of the filter is heated to low redness in a platinum crucible.

The contents are, when cold, moistened with a saturated solution of $(NH_4)_2CO_3$, and evaporated to complete dryness at a gentle heat. The crucible is heated by moving the lamp backwards and forwards beneath it with the hand, and when cold it is weighed.

The filtrate and washings from the calcium oxalate are concentrated in an evaporating dish until NH_4Cl begins to crystallise out. This evaporation may be set going as soon as a sufficient quantity of filtrate has been obtained. The contents of the dish are washed into a flask, concentrated pure HNO_3 is added, and the flask is gently heated. A violent reaction at once takes place, and the decomposition of the NH_4Cl is complete when the reaction ceases, and is not repeated on the addition of a few c.c. of HNO_3. The contents of the flask are transferred to a porcelain dish and evaporated to complete dryness. The residue is heated with a little dilute HCl, washed into a small beaker, rendered alkaline with NH_4OH, and boiled. If found milky, and consisting apparently of magnesium hydrate, a few drops of NH_4Cl should be added to dissolve it. A slight white precipitate, separating in flocks on boiling the liquid, and changing to brown on exposure to the air, usually forms. This consists of $Mn_2H_2O_2$, and should be filtered off and disregarded. The solution is rendered strongly alkaline with NH_4OH, and Na_2HPO_4 is then added in excess, and the mixture is stirred. The beaker is then set aside for twelve hours. The precipitate is filtered off and washed with a mixture of 3 parts water and 1 part of a solution of NH_4OH of specific gravity ·96, until a few drops of the filtrate, rendered acid with HNO_3, show no opalescence on the addition of a few drops of $AgNO_3$. The precipitate is dried, separated from the filter, and strongly ignited in a platinum crucible. The filter-paper is ignited on the lid which is weighed with the crucible.

Quantity **No. 2—Determination of the Sulphur.**—Water is added to the weighed quantity in the beaker, and dilute HCl is gradually introduced in small quantities until it is in very slight excess. The contents of the beaker are gently warmed to expel CO_2, and are filtered. After washing, the precipitate is removed together with the paper to a small flask and heated with concentrated pure HNO_3, a small crystal of $KClO_3$ being occasionally added. Care should be taken that these reagents are free from H_2SO_4. The contents of the flask are washed out into a porcelain dish, evaporated to complete dryness, and the insoluble matter is separated, as before described. The filtrate is heated to boiling, $BaCl_2$ is then added in slight excess, and the beaker is set aside for a few hours. The precipitate is filtered off, washed, dried, ignited in platinum crucible, and weighed. The S is calculated to FeS_2.

The H_2SO_4 is determined in the filtrate from the original quantity, after treatment with dilute HCl, in the same manner, and is calculated to $CaSO_4$.

Quantity **No. 3—Determination of the Alkali Metals.**— These are best determined by Professor Lawrence Smith's method of which we give a brief summary. Full details will be found in Crookes' *Select Methods in Chemical Analysis*.

The silicate is fused with its own weight of pure NH_4Cl, and six times its weight of pure $CaCO_3$, prepared by precipitation with $(NH_4)_2CO_3$ from a hot solution of the pure chloride, the precipitated $CaCO_3$ being filtered off, well washed, and dried.

It must be remembered that the silicate forms only a portion of the materials of which we are considering the analysis, and that a considerable quantity of $CaCO_3$ is already present.

This fact should be taken into account when calculating the quantities of NH_4Cl and $CaCO_3$ to be added. Thus in a cement mixture, containing as it would do about 75 per cent. of $CaCO_3$, it would only be necessary to add NH_4Cl to the extent of one-quarter the weight of the substance taken, allowing about three-quarters its weight as $CaCO_3$.

The required quantities of NH_4Cl and $CaCO_3$ having been weighed off are intimately mixed in a small mortar and added to the weighed quantity of the substance in the crucible, the contents of which are thoroughly mixed by stirring them with a small glass rod or a spatula. The mixture is then heated, at first gently, and finally to full redness for an hour over a blow-pipe flame. When cool the crucible with its contents is placed in a platinum or porcelain dish, water is added, and the dish is slowly heated. The fused mass after a time becomes detached from the crucible and disintegrates, the crucible is then washed into the dish. When completely disintegrated, the contents of the dish are filtered, and the precipitate is well washed. The precipitate is rejected, the filtrate contains the alkalies together with lime and also magnesia, if present, as it almost invariably is in limestone. The liquid is heated, $(NH_4)_2CO_3$ and NH_4OH are added in excess together with a few drops of $(NH_4)_2C_2O_4$, and filtered. The filtrate and washings are concentrated until NH_4Cl begins to crystallise out. The contents of the dish are then washed into a flask of ample capacity, and strong pure HNO_3 is added in sufficient quantity to decompose the ammonium salts present. The flask is heated until the reaction is at an end, the contents are then washed into a porcelain dish, and evaporated to complete dryness. The residue is boiled with baryta-water to remove magnesia, and filtered; the excess of baryta being removed from filtrate by heating with NH_4OH and $(NH_4)_2CO_3$. The liquid is then filtered, and the filtrate is evaporated to complete dryness, after the addition of a few drops of dilute HCl, in a weighed platinum dish. The dish is carefully heated on the sand bath to ensure perfect dryness, and when cold it is quickly weighed. The mixed chlorides are separated as follows:—

Separation of the Chlorides.—The contents of the dish are dissolved in water, a few drops of dilute HCl are added, together with $PtCl_4$ solution in excess. The dish is heated on the water-bath until the contents are in a semi-fluid pasty condition, a little alcohol is then added, and the dish is gently shaken to mix the contents. The double salt crystallises out, as the liquid concentrates on the water-bath. If the liquid becomes colourless or of a very pale straw colour, sufficient $PtCl_4$ solution has not been added. When the precipitate has settled, the supernatant liquid is decanted off and passed through a filter that has been previously washed, dried in the steam-oven, and weighed. The treatment with alcohol is repeated once, the precipitate is then washed from the dish upon a filter with a jet of alcohol, contained in a small wash-bottle, the washing being continued until the washings are colourless. The paper with the precipitate is dried in the steam-oven and weighed. The alkalies are calculated to oxides.

Quantity No. 4—Determination of Carbon Dioxide.—This should be determined by the increase in weight of absorption tubes, and one of the best forms of apparatus for the purpose is that given by Fresenius (*Quantitative Analysis*, vol. i., p. 340), which may be referred to. If FeS_2 has been proved to be absent, the contents of evolution flask may be washed into a platinum dish, evaporated to dryness, the insoluble matter separated and filtered off, and the H_2SO_4 determined in the filtrate.

The Analysis of Clay.

General Classification.—For the purpose of analysis we may divide clays into three classes—

1. Those in which calcium carbonate is absent, or in which it occurs only in moderate quantity, up to say 6 or 7 per cent.

2. Those in which calcium carbonate is present in quantities as large as from 20 to 50 per cent., as, for instance, in gault clay, and in the clays and shales of the lias formation.

3. Alluvial mud from the banks of tidal rivers, such as Medway mud, in which calcium carbonate is either absent or present only in small quantity, but which contains chlorides and sulphates of magnesium, and the alkali metals.

The general method of analysis requires modification in each case.

1. Analysis of Clay Containing Calcium Carbonate in Small Quantity.

Initial Preparation.—The sample in the rough state is dried for some time in the steam-oven. It is then finely powdered, further dried in the steam-oven for several hours, and preserved in a well-stoppered glass tube.

K_2O, Na_2O, FeS_2, $CaSO_4$, and CO_2 are determined exactly as in the analysis of limestone.

Determination of Sand, SiO_2, Fe_2O_3, Al_2O_3, CaO, MgO.— From 1 to 1·5 gramme of the dry clay is weighed from the tube and transferred to a platinum dish. H_2SO_4 diluted with its own volume of water is added, the contents of the dish are heated for about six hours, and are then evaporated to complete dryness. The residue when cold is treated with HCl, the sand and SiO_2 are filtered off, and are washed, dried, ignited, and weighed, and the sand is separated as in the analysis of limestone. The nature of the sand may be determined as follows :— It is ignited and weighed in a capacious platinum crucible, six times its weight of fusion mixture is added to, and stirred with the sand in the crucible, and the mixture is fused until effervescence ceases. The fused substance is treated in the crucible with dilute HCl, heated gently to expel CO_2, evaporated to complete dryness, heated in air-oven for two hours at 150° C., and the SiO_2 is separated and filtered off as previously described. Bases, if present—that is, if the sand is not pure quartz sand— are determined in the filtrate. The Al_2O_3, Fe_2O_3, CaO, and MgO are estimated, as in the analysis of limestone, in the filtrate from sand and SiO_2.

Determination of Combined Water and Organic Matter. —Two sufficiently accurate methods are available according to circumstances, viz. :—

(a) A weighed quantity, from 1 to 1·5 gramme of the clay, after drying in the steam-oven, is ignited over a blast-lamp until the weight is constant. If FeS_2 and $CaCO_3$ are absent the loss may be set down as water and organic matter. If FeS_2 is present the loss consists of water, organic matter, and sulphur, the latter being replaced by oxygen, the iron, previously in combination with it, existing now as Fe_2O_3. The loss per cent. on ignition requires, therefore, to be diminished by the percentage of FeS_2 in the clay, and increased by the equivalent quantity of Fe_2O_3. If $CaCO_3$ is present the percentage of CO_2 must, of course, be deducted from the total per cent. of loss. If FeS_2 and $CaCO_3$ are both present the method (b) should be adopted in preference.

(b) From 1 to 2 grammes of the dried clay are treated in a beaker or evaporating dish with water and very dilute HCl in very slight excess, warmed gently to expel CO_2, and filtered through an ashless filter-paper, which has been previously dried in the steam-oven, and weighed. The residue on the filter is well washed, dried with the paper in the steam-oven for two hours, and weighed. It is then strongly ignited in a platinum crucible, and again weighed. From the loss of weight after ignition, less the weight of filter-paper, the loss per cent. is calculated. This should be corrected, as described under method (a), if FeS_2 is present.

2. Analysis of Clay Containing Calcium Carbonate in Considerable Quantity.

Preparation.—The sample is prepared for analysis as in 1, and K_2O, Na_2O, and CO_2 are determined as in the analysis of limestone.

From 7 to 10 grammes of the dried clay are weighed off, transferred to a beaker or evaporating dish, water is added, also dilute HCl in very slight excess, the mixture is warmed gently to expel CO_2, filtered through a dried and weighed filter-paper, well washed, dried in the steam-oven, and weighed. The weight, less that of the filter-paper, is that of the actual clay in the quantity taken. The filtrate is made up to 500 c.c., 200 c.c. are taken for determination of Fe_2O_3, Al_2O_3, CaO, and MgO, and 200 c.c. for $CaSO_4$ if present. It is concentrated considerably, heated to boiling, a few drops of HCl and of $BaCl_2$ are added, the beaker is set aside, and if a precipitate forms it is filtered off, dried, ignited, and weighed.

The dried clay is removed as far as possible from the filter-paper, finely powdered, and transferred to a well-stoppered tube.

From 1 to 1·5 gramme is weighed off for estimation of sand, SiO_2, Fe_2O_3, Al_2O_3, CaO, and MgO, exactly as described in 1. From 1 to 1·5 gramme is taken for determination of FeS_2, as in 1, if present, and about 1 gramme for water and organic matter.

About 1 gramme is taken for determination of the alkali metals by Lawrence Smith's method, before described.

The constituents are calculated to per cents. of the whole clay.

3. Analysis of Medway or Alluvial Mud.

Preparation.—This material is prepared for analysis as in the previous instances.

About 8 grammes of the dried clay are weighed off, digested with water, filtered through a weighed filter-paper, well washed, dried in the steam-oven, and weighed. If $CaCO_3$ is present, dilute HNO_3 should be used, in addition to the water, in very slight excess. The filtrate is made up to 1 litre.

The clay is separated from the filter-paper, finely powdered, and preserved in a tube. Sand, SiO_2, Fe_2O_3, Al_2O_3, CaO, MgO, FeS_2, Na_2O, K_2O, water, and organic matter are estimated in portions of this, as in the case of No. 2. CO_2, if present, is determined in a proportion of the original clay.

250 c.c. of the filtrate from the clay are taken for the determination of Fe_2O_3, Al_2O_3, CaO, and MgO.

250 c.c. are taken for the determination of the alkali metals.

The quantity is transferred to a beaker, **rendered** alkaline with NH_4OH, heated to boiling, $(NH_4)_2CO_3$ is **added to** remove lime, if present, and filtered. The filtrate is **evaporated** to dryness in a platinum dish, and gently ignited to remove ammonium salts. The residue is boiled with a little baryta-water, filtered, the **excess of** baryta being removed from the **filtrate** with NH_4OH and $(NH_4)_2CO_2$, again filtered, and the filtrate evaporated to **complete** dryness in a weighed platinum dish. After being weighed the mixed chlorides are separated with $PtCl_4$, as described in the analysis of limestone.

Sulphuric acid is determined in 200 c.c. of the solution. 200 c.c., if acid, are rendered neutral or very faintly alkaline with Na_2CO_3, a few drops of K_2CrO_4 solution are added, and the chlorine determined by titration with standard silver nitrate.

The Analysis of Portland Cement and Hydraulic Lime.

Preparation.—The sample, especially in the case of cement, must be reduced to an extremely fine state of sub-division and preserved in a well-closed bottle.

Sand—sometimes described as insoluble residue—SiO_2, Fe_2O_3, Al_2O_3, CaO, and MgO, are determined as in limestone, in weighed quantities of from 1 to 1·5 gramme of the substance, previous ignition being, of course, unnecessary. The silica separates completely on digesting with HCl. Before adding this acid, the weighed quantity should be triturated with a little water until all tendency to set has disappeared.

The alkali metals are determined in about 1 gramme of the substance by Lawrence Smith's method, as before.

About 1·5 gramme is taken for the estimation of H_2SO_4. It is digested with HCl, evaporated to dryness, the SiO_2 separated in the usual manner, filtered off, and the H_2SO_4 precipitated in the filtrate with $BaCl_2$.

Determination of the Sulphur.—About 10 grammes of the substance are taken for the determination of the sulphur, existing combined with calcium as CaS. The weighed quantity is triturated in a porcelain dish with water to a thin slip, until it shows no further tendency to set. It is washed into a flask, fitted with a doubly-bored rubber cork. A funnel-tube passes through one of the apertures and reaches nearly to the bottom of the flask, one limb of a tube, bent twice at right angles, is inserted in the other aperture, reaching only just through the cork. The other limb dips into a solution of cadmium chloride, contained in a tall narrow cylinder or beaker. It should reach nearly to the bottom of the vessel. HCl is passed through the funnel into the flask, which is gently heated until the cement is entirely decomposed, the contents being finally boiled. The contents of the beaker are passed through a filter that has been previously washed, dried in the steam-oven, and weighed. This filter, together with the precipitate, is well washed, dried in the steam-oven, and weighed; the percentage of CaS being calculated from the weight of the cadmium sulphide.

Determination of Carbon Dioxide and Water.—CO_2 and water are determined by the loss on strong ignition in a platinum crucible, of a weighed portion of the cement. CO_2 alone may be determined by the apparatus devised by Dr. Fresenius, previously referred to. By a modification of this apparatus as used for the analysis of black ash (Fresenius : *Quantitative Analysis*, vol. ii., p. 234), both CO_2, and S existing as sulphide, may be determined at one and the same operation.

CHAPTER XIII.

THE EMPLOYMENT OF SLAGS FOR CEMENT MAKING.

Slags sometimes Used for Adulteration.—The days have now gone by, we may hope, when slags of all kinds are regarded with suspicion by cement users, owing to their occasional employment for the **adulteration of** Portland cement. It cannot be denied that for this purpose certain varieties of iron slag were extremely well adapted, since they approximate so closely in colour, weight, and composition to the cement itself that their presence could be detected only with difficulty, even by means of delicate chemical tests.

Nature of Slags Used to Mix with Cement.—The slags used for fraudulent additions to cement were generally those which had a tendency to disintegrate on exposure to the atmosphere, and which go abroad naturally into a bulky gray powder, having at times an astonishing resemblance to Portland cement. The fact of this spontaneous disintegration would seem to argue that some change, either of a physical or chemical nature, took place in consequence of atmospheric action; probably these varieties of slag were selected chiefly because a powder could be obtained from them by mere sifting, without the expense and trouble of grinding, which is no easy matter in the case of the dense inert slags.

Some Slags Unacted upon by Moisture or by the Atmosphere.—Certain varieties of slag would seem to be very little affected by air or moisture, and remain for years on the spoil-banks with arrises as sharp and clean as they were upon the day when they were tipped. Doubtless it was the apparent inertness of these materials which led experimenters to form the opinion that it was impossible to use them in cement-making, for the great similarity in composition of some descriptions of blast furnace slag with the clinker of Portland cement must have frequently caused attention to be directed to this material, in order to find a possible use for it as a cement component.

Slags Produced at Extreme Temperatures.—It must here be remembered, however, that the temperature at which slag is produced is greatly in excess of that needed to prepare Portland cement clinker, and, moreover, that in the case of the clinker too high a temperature in the kiln, resulting in complete fusion of the mass, leads to the formation of a blue-black glassy substance (slag), which in consequence of its vitrification is of no practical use for cement. The analogy of the two processes would again deter the investigator from further attempts to employ slag in cement-making.

Composition of Blast-furnace Slags.—A glance at the annexed table will show that many iron slags contain all the essential ingredients of Portland cement, though in none of them are they present in the same proportions as those selected by the cement maker. In almost every case there is a marked deficiency of lime, and nearly all the attempts made to utilise slag have involved the addition of lime in some form or other to the slag in a finely-divided state.

ANALYSES OF VARIOUS SLAGS* FROM IRON WORKS.

Locality.	Silica.	Alumina.	Lime.	Iron, Alkalies, Sulphur, &c.
1. Staffordshire, . .	41·99	23·27	31·62	3·21
2. South Wales, . .	42·84	28·84	26·13	2·11
3. ,,	32·30	15·70	44·80	7·20
4. Yorkshire, . .	39·47	23·66	32·12	4·75
5. Lancashire, . .	33·49	10·12	46·97	8·54
6. Staffordshire, . .	49·05	10·84	34·33	5·78
7. Warwickshire, .	41·30	16·20	40·80	2·70
8. ,, .	36·30	13·86	30·49	9·76
9. Scotland, . . .	32·10	24·28	35·43	8·10
10. Nottinghamshire, .	32·18	31·50	33·05	3·27
11. Lancashire, . .	38·68	13·13	45·82	2·37
12. Cumberland, . .	34·00	15·10	47·20	3·70
13. Derbyshire, . .	39·24	23·04	32·06	5·66
14. ,, .	40·00	11·50	40·60	7·90
15. Lincolnshire, . .	32·86	21·88	40·34	6·12
16. ,, . .	31·37	19·69	40·23	8·71
17. Northamptonshire, .	39·06	21·54	37·16	2·24
18. North Wales, . .	31·28	12·41	46·10	10·21
19. Newcastle, . .	26·37	14·42	48·84	10·07
20. ,, . .	34·15	16·68	43·39	5·78
21. ,, . .	31·05	23·15	36·51	9·29
22. Cleveland, . .	32·15	17·53	45·50	4·82
23. ,, . .	35·45	21·55	33·70	9·30
24. ,, . .	29·89	26·13	34·35	8·63
25. ,, . .	30·98	28·03	33·33	7·60

* A great number of these are of no use for cement making.

Some Former Attempts to employ **Slags for Cement-Making.**—We cannot now afford sufficient space to record the various directions in which the problem of slag utilisation for cement production has been approached. Some have thought that by heating the requisite quantity of lime with the powdered slag a partial combination between the silica compounds and the added lime might be effected. Others have endeavoured to bring about this combination in the wet way by grinding slag and lime together in a mortar-mill, but all these processes have hitherto proved unsatisfactory when carried out upon a commercial scale.

Process of Messrs. Bosse & Wolters.—The only successful process of making a cement resembling Portland cement from slag, which has come under our notice, is that patented by Messrs. Bosse & Wolters, whose representative in this country is Mr. E. Larsen. Their invention consists in the selection of suitable slags, which are mechanically reduced to an extremely fine powder, and amalgamated in a machine of simple construction with a proportion of slaked lime. In the ordinary process 25 parts of slaked lime are added to 75 parts of powdered slag. The secret of success lies in the extremely fine grinding and the perfect amalgamation of the slag and lime. In carrying out the manufacture, the slag, as it issues from the furnace, is run into water, giving rise to the formation of so-called "slag sand." This material was patented many years ago by Mr. Charles Wood, of Middlesbrough. It is found that when the molten slag falls into water it behaves much in the same way that glass does when used to produce Prince Rupert's drops, it flies into countless sharp fragments, resembling sand, the individual portions, as they fall, take the form of a porous pumice-like mass, which is very friable and capable of being easily crushed into powder. Slag sand produced in this way has, however, certain disadvantages for the after processes to which it is subjected; it is, moreover, extremely retentive of water, and difficult to dry.

Process of Mr. Snelus.—Mr. G. J. Snelus, F.R.S., has described a process of making cement from slag, in which the moisture is expelled from the granulated slag by calcining it in a furnace, and he states that in the course of this process a notable quantity of sulphur, which is always present in slags as calcium sulphide, is gradually oxidised to calcium sulphate, accompanied by the evolution of sulphuretted hydrogen. The calcium sulphate, if present in moderation, beyond as we have seen tending to render the cement slow in its set, has no injurious action. The process of Mr. Snelus, which he has perfected, in conjunction with Mr. J. Gibb, does not involve the use of additional lime beyond that originally present in the slag. The following results in pounds per square inch were obtained on testing three samples of this cement neat:—

	7 days old.	28 days old.	3 months old.
No. 1.	420	470	600
No. 2.	370	560	650
No. 3.	390	530	720

We are not in possession of full details respecting the process of manufacture, nor have we an analysis of this cement, but Mr. Snelus stated in 1890 at the Society of Arts, London, that 100 tons of slag were being used weekly for this purpose.

Mode in which the Slag Sand is Prepared.—There can be no doubt, however, that the act of forming the slag sand in the manner above described has a considerable influence on the behaviour of the material in the presence of lime, for it is found that when slag is allowed to cool slowly in the ordinary way, and is then reduced by mechanical means to a fine powder, it no longer possesses hydraulic properties in conjunction with the added lime. It is, therefore, probable that the rapid cooling or "chilling" has some physical effect upon the ultimate molecules of the slag, causing them to assume a state in which they can more readily undergo chemical action. At the time of entering the water, and in the fierce ebullition which then takes place, there is, moreover, undoubtedly a certain amount of chemical action, for with the steam an appreciable volume of sulphuretted hydrogen gas is given off, due probably to the decomposition of the calcium sulphide present in the slag.

Mode of Manufacturing Slag Cement.—In preparing the cement under Messrs. Bosse & Wolters' specification, the slag sand is first thoroughly dried, and to 3 parts by weight of the dry slag sand 1 part of dry slaked lime is added in an apparatus termed by them the "homogenizer," but better known as the "ball-drum." As will be seen from Fig. 11, this consists of a hollow drum or cylinder, which may be of various sizes, the interior circumference is lined with fluted cast-iron plates, and in this drum are placed a number of chilled iron or steel balls, about 1 inch or 1½ inches in diameter. The mixture of slag and lime is passed into the drum through a hollow trunnion, by means of a worm feeding apparatus, and when a full charge has been added, the aperture is closed and the cylinder is caused to rotate slowly on its axis with the result that the contents are, by the constant collisions of the revolving balls on the fluted lining, pounded and reduced to an extremely fine state of subdivision, and at the same time most perfectly mixed. When this operation has been continued for an hour, or for such time as may be found sufficient to thoroughly grind the mixture, a door in the outer lining is opened and the contents in the course of a few revolutions are ejected through a shoot into a bag or barrel ; the balls being retained by a coarse grating. The charge thus produced is now ready for the market, not requiring to be sifted or handled in any way.

Manufacture is Economical.—It will be seen from this description that the manufacture of slag cement is an extremely simple one, needing but little skilled labour and a relatively cheap plant. The area needed for the works is small, and a French writer, who has had considerable experience, states that the entire cost of works and plant for an annual output of 6,000 tons (say 20 tons of cement per diem) is under £6,000, in which estimate he includes all the necessary buildings and an engine of 150 horse-power. This contrasts very favourably with the cost of works to produce an equivalent quantity of Portland cement by any of the existing methods, as will be evident from our estimates in Appendix F, p. 218.

Features of importance in the Manufacture.—There can be no doubt that this process owes its success—first, to the use of slag sand, which, as we have seen, possesses certain properties in which ordinary ground or pounded slag is deficient ; and second, to the extremely fine grinding secured by the use of the homogenizer, which undoubtedly produces a more impalpable powder than any other similar machine with which we are acquainted. It must be remembered that in slaking the lime, if this operation is carefully conducted, with only just the proper amount of hot water, we obtain the quicklime, which constitutes 25 per cent. of the bulk of the cement, in a state of subdivision infinitely more minute than would be possible by any mechanical means, and the pumice-like friable slag sand lends itself admirably to the grinding process, and has its pores mechanically filled with the lime particles in the homogenizer. Experiments tend to show that whereas ordinary slag is unacted upon by water, it is possible by grinding it to a very fine powder in an agate mortar to obtain a substance which has feeble cementitious qualities ; this is probably due to the liberation of particles of lime and other sparingly soluble bases which have become surrounded in the fused mass with a thin film of silicates.

Necessity for Extremely fine Grinding.—The urgent need of extremely fine grinding is shown by some tests at Choindez, which gave the following results :—

Meshes per square inch.	Percentage Residues.			Age in Days.	1 Cement to 3 Sand.		
	A.	B.	C.		A.	B.	C.
5,806	7·1	0·5	0·0	...	...	...	...
16,128	16·1	2·0	0·5	7	131	227	341
32,257	28·9	18·0	8·4	28	220	419	537

In this table A, B, and C represent the same sample of cement ground in three different ways, and tested by sieves of three differing degrees of fineness. Sample A was the most coarsely ground, in that on a very fine lawn it left a residue of 28·9 per cent. On a corresponding sieve, the more carefully ground sample C left a residue of only 8·4 per cent. The sample A was ground as fine as it was possible by means of the mill-stones in common use in cement works. Sample B was passed through the homogenizer, and sample C was treated a second time in the homogenizer, so as to be reduced to the finest possible powder. The cement C, when tested with three parts of sand, proved to be more than double the strength of sample A, tested in a similar way. Portland cement can, of course, be dealt with by the homogenizer and gains greatly in tensile strength by the operation, but the improvement is chiefly noticeable when the cement is tested with sand. It will at once be seen that the finer the grinding the greater is the power of uniting together the sand particles, as the action of a cement depends to a very great extent upon the amount of surface that it will cover.

Properties of Slag Cement.—We may now glance at the properties of the slag cement, which differs to a marked degree from Portland, for while the specific gravity of a good sample of the latter would average about 3·10, that of slag cement never exceeds 2·73, the slag from which it is made having a specific gravity of about 3·00, and that of slaked lime being only 2·08, the diminished density being no doubt partly due to the porosity of the slag sand. A cubic foot of the slag cement weighs, therefore, about 75 lbs., or the bushel would weigh 90 lbs., being thus about 20 per cent. lighter than Portland. Slag cements are essentially of the slow setting type, rarely taking less than five hours for this purpose. They distinguish themselves from Portland chiefly from the fact that they reach their greatest degree of induration in less than twelve months; in fact, in many cases they show but little improvement after the first month. They test extremely well with sand, owing to the extremely fine grinding, and there is less difference between the tests of the neat cement and that with 3 parts of sand than is observed in the case of most samples of Portland cement. Owing to the slowness of the set they are ill-adapted for plastering and stucco, and they do not produce so hard a surface as that obtained by the use of Portland.

Slag Cement can be Stored with Impunity.—Slag cement compares favourably with other similar materials when stored for long periods. If freely exposed to the air, the lime gradually unites with carbonic acid, and this action is relatively much more rapid than is the case with Portland , thus in four days a thin layer of slag cement gained about 5 per cent. of carbonic acid, the amount of this acid taken up in the same time by a similar sample of Portland cement was only 1 per cent. The slag cement does not, however, greatly deteriorate in tensile strength by prolonged storage, as the following experiment tends to prove :—A sample of the cement was tested with sand, and gave, at seven days, 254 lbs., and at 28 days 340 lbs. per square inch, with 3 parts of normal sand. After being kept for fifteen months, and similarly tested, the figures were 243 lbs. at seven days, and 321 lbs. at twenty-eight days, showing, therefore, a loss in strength of less than 5 per cent. Slag cement yields such a fat rich mortar that it lends itself peculiarly well for works under water. Used for sea-walls, as has been done at Skinningrove, near **Saltburn-on-the-Sea**, it rapidly attains great hardness, and makes excellent work.

The Chemistry of Slag Cement.—The chemistry of the cement action in the case of this material is somewhat obscure, and tends to subvert certain of the theories which have been put forward to explain the induration of Portland cement. Before considering the question further we may glance at the following analyses :—

	Portland Cement.	Slag Cement.
Lime (CaO), ...	61·07	39·68
Silica, ...	21·70	24·34
Alumina, ...	8·60	8·74
Protoxide of iron, ...	...	0·27
,, manganese, . . .	...	0·23
Peroxide of iron, ...	2·25	0·14
Magnesia, ...	1·17	6·59
Potash, ...	0·75	0·28
Soda, ...	0·42	0·44
Sulphur, ...	...	0·93
Sulphuric acid, ...	1·93	0·25
Carbonic acid, ...	0·80	4·07*
Water and loss, ...	1·26	4·70
Total, . . .	99·95	100·66

* Carbonate of lime equal to lime 2·28, carbonic acid 1·79.

Here we have in the first column a sample of Portland cement of average composition, and in the second an analysis of slag cement. If the carbonate of lime in the latter analysis is separately set down as lime and carbonic acid we have in the slag cement $39.68 + 2.28 = 41.96$ per cent. of lime, as compared with 61.07 per cent. in the Portland cement, though probably the 6.59 per cent. of magnesia in the former may be regarded as ranking with the lime.

Proportion of Lime and Magnesia to Acids.—If we take the total percentage of lime and magnesia in both cements, equal to 48.55 in the slag cement and 62.24 in the Portland, we find that the equation of total lime divided by silica + alumina is roughly $\frac{3}{4}$ in the slag cement, as compared with $\frac{2}{1}$ in the Portland. This latter equation is fairly representative of the average composition of all good specimens of Portland cement, and we should at first sight consider it quite impossible that a cement having the formula of $\frac{3}{4}$ would give good results.

Composition of Slag Cement throws some doubt on Chemistry of Portland.—We are in fact compelled to reconsider certain of our theories respecting the action of the silicates and the relative proportion of lime required to obtain cementitious properties. It is probable that in the case of slag cement, at the very high temperature attained in the blast furnace (that of complete fusion), compound silicates of lime and alumina, together with small quantities of metallic and alkaline silicates, are produced which, when taken alone, are almost entirely inert. That is to say, that when they are reduced mechanically to a fine state of subdivision there is but little tendency among the particles to form fresh compounds in the presence of water. The energetic action of hydraulic cements of the Portland type is largely due, as we have elsewhere shown, to the formation of hydrated silicates of lime, which rapidly solidify the water used in making the mortar, and are said to set. In nearly all specimens of Portland there is undoubtedly a considerable margin of free or uncombined lime. This lime may either have been present in the original mixture in excess of the requirements of the silicic acid, or it may have become mechanically entangled in or coated with the vitreous, or semi-vitreous, silicates formed in the kiln. It is always found in Portland, to which it gives its alkaline reaction, and it may be dissolved out from the cement in considerable quantities by a saccharine solution.

Free Lime in Portland Cement.—It is difficult to estimate the exact proportion of this free lime, but we are warranted by what takes place in the case of slag cements to consider that the amount is considerably larger than chemists have hitherto thought possible. When **Portland** has been steamed or thoroughly "purged," and the whole of the available caustic lime has been hydrated, the cement loses its energy of set, and is rendered much safer and more reliable in its action. What takes place, when water is added to the mixture of pulverised slag and hydrate of lime in the slag cement, must be a partial decomposition of the sparingly soluble silicates and a mutual reaction between them and the lime, which latter in itself (being fully hydrated) is incapable of further chemical action in the presence of water. The slag contains but little free lime, as it is but slightly attacked by acids, but there is no doubt a slow and gradual decomposition of the silicates of the alkalies and the other bases present in the slag, and a reaction ensues between these silicates and the calcium hydrate, leading to the formation of fresh compounds of lime, silica, and alumina, which are not only themselves capable of induration, but possess also the power of uniting very considerable quantities of aggregates in the form of mortar and concrete.

Excess of Lime in Portland Cement.—If this theory is accurate, it will be seen that it follows as a necessary consequence, either that in Portland cement the lime is present in dangerous abundance (to the extent of something approaching 20 per cent.), or that in the slag cement, in consequence of the high temperature of calcination, different combinations of the lime and other substances are obtained, which would, however, appear to be quite as stable and permanent as those prepared at a much lower temperature in the cement kiln.

Theory of Cement Action Needs Revision.—We cannot, however, escape the conviction that the behaviour of the mechanical mixture of silicates and calcium hydrate forming the slag cement throws considerable doubt upon those theories of cement action which imply that in Portland cement the induration is brought about by the hydration of ready formed silicates, or that certain silicates of lime, alumina, iron, and the alkalies, produced in the kiln by the action of heat, are in the presence of water recombined or transformed into hydrated double silicates, since it is evident that in the slag cement notable quantities of hitherto uncombined lime are united to the silicates of alumina or other bases present in the slag. The exact nature of the chemical changes which take place has, we regret to say, not yet been satisfactorily explained.

Other Considerations involved besides Chemical ones.—That other considerations, besides purely chemical ones, influence the behaviour of the slag is, however, shown by the following samples of slag sand, which when analysed gave almost identical results, but which, when treated in the same way and mixed with precisely the same amount of lime, gave widely different results. The composition of the slags was as follows :—

Ingredients.	A.	B.
Silica, ...	24·10	23·22
Alumina, ...	16·30	15·61
Lime, ...	46·53	47·10
Oxide of iron, ...	0·93	0·78
Soda, potash, &c., . .	5·04	5·56
Carbonic acid, . . .	0·65	0·85
Moisture, ...	6·45	6·88
Totals, . .	100·00	100·00

Tested with sand in the same way, the tensile strength of the cements produced by the addition of lime was—

1 Cement to 3 Sand.			
A.		B.	
7 days.	28 days.	7 days.	28 days.
Lbs.	Lbs.	Lbs.	Lbs.
326	431	150	208

showing a difference of over 50 per cent.

German Tests of Slag Cement.—In order to show the behaviour of this cement when tested, the author considers that the following details of the examination of a sample made and sealed up in his presence and forwarded to the Imperial German Testing Station at Berlin in December, 1886, may be of interest, and it will serve also to show the very careful and searching manner in which such trials are conducted, and speaks well for the qualities of this new material. The investigation is extracted from the *Proceedings of the Institution of Civil Engineers :—*

INVESTIGATION OF SAMPLE OF SLAG CEMENT CARRIED OUT BY THE ROYAL TESTING-ESTABLISHMENT FOR BUILDING MATERIALS AT BERLIN.

Results of the examination of a parcel of cement forwarded on the 3rd December, 1886, marked "Sample of Cement," without further particulars as to locality. The tests were commenced on the 9th December, 1886, under the rotation number Spec. XVI., No. 4,764.

As a mean of three observations in each case :—

	Kilogrammes.	Lbs. per Cubic Foot.
1 litre of the cement lightly filled = . .	1·042	65
„ „ shaked down = . .	1·611	100·5
„ normal sand = ...	1·640	102·3

The normal sand was obtained by the selection of the finest possible quartz sand, which was washed and dried. From it were removed all the coarse particles rejected by a sieve of 60 meshes to the square centimetre, and only the particles subsequently retained on a sieve of 120 meshes per square centimetre were employed.

The production of a suitable mortar needed 40 per cent. of water, 500 grammes of the cement being used. A slightly stiffer mortar required 36 per cent. of water; the temperature of the cement and the water being the same, and equal to that of the atmosphere of the laboratory, the rise of temperature on mixture was 0·7° Centigrade. It was made up into pats on glass, and set in $2\frac{3}{4}$ hours, the average temperature throughout the experiment being 22·5° Centigrade.

Tested with the sieve, the following results were obtained :—

       5,000 meshes per square centimetre, 14·0 per cent. residue.
           900   „   „   „   5·0   „   „
           600   „   „   „   1·0   „   „
           324   „   „   „   0·5   „   „
           180   „   „   „   0·0   „   „

The tests for expansion and contraction were carried out by means of ten pats of neat cement, made up on glass and roofing-tile, and trowelled to a thin edge. The pats remained twenty-four hours in air, and were then placed in water. They kept perfectly true and sharp-edged, and free from cracks. There was no expansion or contraction, and the cement adhered to the glass. When broken it presented a close even-grained and uniform fracture.

The briquettes employed to ascertain the tensile and compressive strength were made up on December 10, 1886, as follows :—

(a) The neat cement was mixed with $16\frac{1}{2}$ per cent. of water.

(b) The sand briquettes were made of 1 part by weight of cement, 3 parts by weight of normal sand, and $7\frac{1}{2}$ per cent. of water. The mixture was beaten into metal moulds placed on glass plates in the usual way.

The temperature of the air was 22·5° Centigrade.
,,   ,,   ,,   water was 16·2° Centigrade.
The moisture of the air was 68 per cent.

The tests set the first day in air, covered up with writing paper to avoid too rapid desiccation ; the remainder of the time they were in water. Taken from the water and tested they gave the following results :—

Mix-ture.	Close-ness after Filling	TENSION.				Close-ness after Filling	COMPRESSION.				Proportion of Tensile to Compressive Strength.	
		Tensile Strength after					Compressive Strength after					
		7 days.		28 days.			7 days.		28 days.		7 days.	28 days.
		Ger-man.	Eng-lish.	Ger-man.	Eng-lish.		Ger-man.	Eng-lish.	Ger-man.	Eng-lish.		
		Kils. per sq. cm.	Lbs. per sq. in.	Kils. per sq. cm.	Lbs. per sq. in.		Kils. per sq. cm.	Lbs. per sq. in.	Kils. per sq. cm.	Lbs. per sq. in.		
Neat. 1 to 0	2·113	45·50	647·0	48·65	692·0	...	...	...	...	...	...	...
Sand. 1 to 3	2·261	30·03	427·0	35·78	509·0	2·262	237·3	3,376	300·1	4,269	1 to 7·902	1 to 8·337

NEAT CEMENT.

No.	7 days.		28 days.		Remarks.
	German.	English.	German.	English.	
	Kils. per sq. cm.	Lbs. per sq. inch.	Kils. per sq. cm.	Lbs. per sq. inch.	
1	44·25	629	48·25	686	Made on metal plates, with 16½ per cent. of water.
2	45·50	642	45·75	651	
3	43·25	615	53·00	754	
4	48·00	683	49·00	697	
5	44·00	625	48·00	683	
6	46·50	653	48·50	694	
7	45·75	651	48·25	686	
8	44·25	629	52·50	746	
9	48·00	683	46·25	657	
10	45·50	647	47·00	668	
Total,	455·00	6,457	486·50	6,922	
Average,	45·50	646	48·65	692	

1 Part Cement and 3 Parts Sand.

No.	7 days.		28 days.		Remarks.
	German.	English.	German.	English.	
	Kils. per sq. cm.	Lbs. per sq. inch.	Kilos. per sq. cm.	Lbs. per sq. inch.	
1	31·50	448	38·25	544	Made on non-absorbent plates, with 7½ per cent. of water.
2	30·00	427	35·00	498	
3	28·50	405	36·50	518	
4	31·50	448	37·75	536	
5	29·00	412	35·50	505	
6	31·00	440	34·50	490	
7	30·50	433	35·75	508	
8	31·50	448	34·50	505	
9	27·25	387	33·50	476	
10	30·00	430	36·50	519	
Total, .	300·75	4,278	357·75	5,099	
Average,	30·07	427	35·78	509	

The briquettes were kept for the first day in the air, and the remaining time in water.

1 Part Cement and 3 Parts Sand.

No.	7 days.		28 days.		Remarks.
	German.	English.	German.	English.	
	Kilos. per sq. cm.	Lbs. per sq. inch.	Kilos. per sq. cm.	Lbs. per sq. inch.	
1	229·6	...	287·8	...	Made on non-absorbent plates, mixed with 7½ per cent. of water.
2	252·0	...	292·3	...	
3	237·4	...	305·8	...	
4	235·2	...	303·5	...	
5	233·0	...	292·3	...	
6	234·1	...	304·6	...	
7	229·6	...	291·2	...	
8	246·4	...	315·8	...	
9	230·7	...	292·3	...	
10	245·3	...	315·8	...	
Total, .	2373·3	...	3001·4	...	
Average,	237·3	3,376	300·1	4,269	

CHAPTER XIV.

SCOTT'S CEMENT, SELENITIC CEMENT, AND CEMENTS PRODUCED FROM SEWAGE SLUDGE, AND THE REFUSE FROM ALKALI WORKS.

Discovery of Scott's Cement.—In the course of some attempts to produce an artificial hydraulic lime by calcining lumps of chalk in a common fireplace, about the year 1854, General (at that time Captain) Scott, R.E., found to his surprise that the calcined lime would not go abroad in the usual way in water. Tested with acid the lime was found to be properly burned, but it had lost its avidity for water. Captain Scott was thoroughly puzzled by this result and he consulted Dr. Faraday in his difficulty.

Faraday's Opinion on the Theory of this Cement.—After careful consideration of the facts that eminent chemist came to the conclusion that this change in the behaviour of the lime was due to the formation of some form of sub-carbonate of lime, a compound the existence of which previous investigators had suspected, and which was believed to confer new properties upon the calcined lime; the most important of these being the above mentioned failure to fall into powder when quenched or sprinkled with water.

First Patent for Scott's Cement.—So convinced was General Scott of the accuracy of this surmise that he was induced to patent the process provisionally in March, 1854, No. 735. In his specification he proposed to prepare a cement either by calcination, so applied as to drive off only a portion of the carbonic acid contained in chalk or limestone, leaving the substance in the state of sub-carbonate, or by subjecting ordinary quicklime or supercalcined lime to heat in the presence of carbonic acid so as to bring it back to the state of a sub-carbonate. He also specified a third process of mixing quicklime and carbonate of lime in such proportions as to cause them to form, when properly treated, a sub-carbonate.

Behaviour of Lime due to Sulphur in the Fuel.—Subsequent experiments proved this invention to be founded on a misconception, and it was ultimately discovered that the change in the behaviour of the lime was due to the presence of small quantities of sulphate of lime, produced from the sulphur in the fuel. When the calcination was carefully carried out, and the resultant quicklime at a red heat remained for a short space of time in contact with brassy coal or impure coke, some of the sulphurous acid was absorbed by the glowing lime, and combined with it to form a sulphite which ultimately passed into a sulphate of lime, and General Scott was the first to ascertain this peculiar action of sulphur compounds and sulphuric acid on quicklime.

Nature of Amended Patent.—General Scott at once perceived the importance of these new discoveries, and he took out a patent for converting lime of a partially hydraulic character into cement by the action of the fumes of burning sulphur. He effected this operation by reheating calcined lump lime in an oven, having a perforated floor, beneath which were placed pots of sulphur, the sulphurous acid from which ascended among the red-hot lime, leading to the formation of calcium sulphite, and this in turn became oxidised into the sulphate. The amount of sulphuric acid thus absorbed by the whole bulk of the lime was small, rarely exceeding from 2 to 3 per cent., and of course only the exterior surfaces of the lumps became coated with the sulphur compound, but .when the cement was ground, to prepare it for use, the sulphate of lime became evenly distributed throughout the mass.

Similar Results obtained in other Ways.—In course of time General Scott found that he could obtain the same results, either by adding sulphuric acid to the water used in preparing the mortar, or by the addition of powdered gypsum or plaster of Paris to the ground lime. It little mattered in what form the sulphuric acid was conveyed to the lime, and many soluble sulphates were found to answer quite as well as the sulphate of lime.

Very small Amount of Sulphur required.—In laboratory experiments very minute quantities of sulphuric acid proved sufficient to control the avidity of caustic hydraulic lime for water. As little as $\frac{1}{4}$ per cent. of calcium sulphate only being needed if used in the water employed for tempering the mortar, while stirring in the lime, previously ground to a fine powder.

Selenitic Cement.—Ultimately General Scott specified the manufacture of a cement, which he named "selenitic cement," by the addition of 5 per cent. of ground plaster of Paris to calcined hydraulic lime, which was then ground to an impalpable powder and placed in sacks or casks for use. The theory of the action of the sulphuric acid was originally supposed to be as follows :—

That each molecule of lime combined with the sulphuric acid and water to form sulphate, and that the sulphuric acid then travelled on to the next molecule of caustic lime, which was in turn converted into a sulphate; the sulphuric acid traversing in this way the whole bulk of the lime, leaving behind it the calcic hydrate which has thus been formed without much evolution of heat and with no apparent change of volume. This action would be a species of cementation, but, in lieu of the above theory, a German chemist, Mr. F. Schott, propounds with much greater show of reason the following explanation :—

Schott's Theory of Selenitic Action.—The particles of gypsum in solution are mechanically deposited over the molecules of lime owing to surface-attraction. The coating of sulphate retards the access of water to the lime; it forms, if we may so style it, a temporary varnish, through which, however, owing to its solubility, the water speedily penetrates, and the molecules of lime then become hydrated, but this action is retarded to such an extent that the combination with water takes place gradually, without much evolution of heat, and with little or no perceptible increase of volume. This latter fact is proved by the much greater density of the selenitic hydrate than that of a hydrate formed in the ordinary way.

The paste formed by the selenitic process is sufficiently bulky to penetrate the interstices between the sand-grains and to bind together a large quantity of this sand into a mortar, though, of course, it is far less finely divided than the particles of fully slaked lime from which mortar is ordinarily made. The hydrate used for common mortar is, however, not really a binding agent in the true sense of the word at all.

Sulphur has no effect on Pure Limes.—If pure lime is treated in this way with a soluble sulphate no retarding influence is exerted, for the hydration takes place instantaneously before the coating is formed; it is only in the case of hydraulic limes, in which the combination with water is more gradual, that the coating of sulphate has time to form. Some interesting experiments by Mr. F. Schott, elucidating this theory, were published in *Dinglers' Polytechnische Journal*, in vol. clix., 1873, p. 30.

Feebly-Hydraulic Limes best adapted for Treatment.— The selenitic process may be employed with advantage with all limes of a feebly hydraulic character, and it also greatly improves the eminently hydraulic limes of the lias formation. Limes of this kind, when made selenitic, will carry a largely increased volume of sand and give good results. Strong tough mortar and good plasterers' stuff may be produced from a mixture of 1 part of

selenitic lime with from 4 to 6 parts of sand, and it is a matter of
common experience that any given sample of lime when prepared by the selenitic process will give double the tensile
strength with twice as much sand, as when slaked and mixed
into mortar in the ordinary way.

The Colour of Selenitic Cement.—In consequence of the
fact that the lime does not become slaked when employed in
accordance with General Scott's process, it retains its original
colour, a warm buff, and the mortar is of course much more
dense than one made from the hydrate. The power of binding
together large volumes of sand is strikingly shown by the tests
to which this material and the original Scott's cement have been
at different times subjected.

Improvement Effected by Selenitic Process.—The improvement effected upon limes, when treated with a small percentage of plaster, as compared with the same materials when
used in the ordinary way, is well seen in the following table,
which gives some tests carried out at the New Law Courts,
under the late Mr. Street, R.A. They were made by Mr. A. W.
Colling, the Clerk of the Works. The resistances were arrived
at by pulling asunder two bricks united crosswise, so as to give
a joint having an area of $18\frac{1}{4}$ square inches. One month was
allowed for setting. In every case the mortar was made in a
mill, and the mean of three tests is indicated :—

Material and mode of preparation.	Proportion of sand to lime.	Mean resistance in lbs.
Lias gray lime as common mortar, .	3 to 1	112
„ „ selenitic „ .	6 „ 1	$209\frac{3}{4}$
Barrow lime as common mortar, .	3 „ 1	125
„ „ selenitic „ .	5 „ 1	$283\frac{3}{4}$
„ „ „ „ .	6 „ 1	196

Tests of Selenitic Cement.—From a large number of tests
made by the author to ascertain the strength of selenitic cement,
as compared with Portland cement, the following have been
selected because they clearly indicate the advantages of the
selenitic treatment when carried out with suitable limes. The
lias lime here used was that from Barrow-on-Soar, and the gray
lime was from the Burham pits on the Medway. The whole of
the samples were prepared by the late Mr. Hartley, who had a
long experience in cement testing. The tests were in all cases
made with ordinary stock bricks bedded across one another at
right angles, giving a joint with an area of 20 square inches.

Nature of Material.	Age in Days when Fractured.	Parts of Sand to 1 of Cement or Lime.			
		3 Sand.	4 Sand.	5 Sand.	6 Sand.
Portland cement,	28	...	463	325	313
Barrow selenitic,	28	541	418	399	399
Burham selenitic,	28	484	454	368	408
Portland cement,	35	...	520	433	309
Barrow selenitic,	35	435	539	438	430
Burham selenitic,	35	424	430	490	556

Kirkaldy's Tests of Selenitic Cement.—By far the most important series of tests of this material with which we are acquainted was carried out for the Selenitic Cement Company by Mr. D. Kirkaldy in 1872. (See Appendix A., p. 206.) He ascertained the resistance of selenitic cement both to forces of tension and compression. In the former set of experiments he made use of test-briquettes having an area at the neck of 5 square inches, and also of bricks bedded crosswise, having a sectional area at the joint of 18·5 square inches. The blocks used for the tests in compression had a base area of 7·84 square inches. The test-briquettes of common lime mortar, when broken eight weeks after making, showed a mean strength of 23·6 lbs. per square inch. Selenitic mortar, made from the same lime and with double the sand, attained in a similar time a strength of 83·0 lbs. per square inch. Blocks of common mortar crushed under a load of 121·7 lbs. per square inch, but when treated selenitically and used with twice the sand, the blocks withstood a load of 629·6 lbs. per square inch.

Experiment in Slaking Lime Powder.—The selenitic action is likened by Mr. Schott, to whose theories we have already alluded, to what takes place when dry, finely-powdered quicklime is tightly packed in a metal cylinder with very minute perforations. If such a cylinder of lime is placed in water, which can then only reach the lime in very small quantities and very gradually, the lime is converted into a dense hydrate, which resembles in every way the hydrate formed by the selenitic process. This experiment would seem to confirm the accuracy of Mr. Schott's theory of the selenitic process.

Mr. Graham Smith's Observations on Sulphates.—It is much more difficult to understand the reason of the influence of sulphate of lime on slaked lime, as observed by Mr. Graham Smith, who, in a paper on "The Effect of Sulphates on Lime Mortar," has shown that calcium sulphate has a notable influence when mixed in certain proportions with slaked hydraulic limes. In 1870, while he was in charge of the various cements and mortars employed in the works in progress at the Liverpool Docks, he carried out a large number of experiments with Halkin lime from Flintshire, a variety of white lias with fair hydraulic properties.

The analysis of this limestone is as follows :—

ANALYSIS OF HALKIN MOUNTAIN LIMESTONE.

Composition.		Centessimally.
Substances soluble in acids— 74·726 per cent.	Carbonate of lime, . .	71·546
	Carbonate of magnesia, .	1·348
	Protocarbonate of iron, . ⎱ Sulphide of iron, . . ⎰	1·040
	Alkalies, ...	0·792
Substances insoluble in acids— 25·274 per cent.	Silicic acid, . . .	20·068
	Alumina, ...	3·521
	Sesquioxide of iron, &c., .	1·192
	Water and carbonaceous matter, . . .	0·493
	Total, . . .	100·000

The whole of the mortar used in the tests was mill-made, ground 30 minutes, the proportions indicated are in all cases by volume. The briquettes were of the ordinary shape, $1\frac{1}{2}'' \times 1\frac{1}{2}''$ or $2\frac{1}{4}$ square inches in area. They were broken in a Michele lever cement-testing machine. Corresponding tests were made with bricks bedded crosswise on the flat, giving a joint of $4\frac{1}{4}'' \times 4\frac{1}{4}'' = 18$ square inches, and the results though similar were scarcely so favourable to the sulphate mortar as the briquette tests which we have appended.

Common Slaked Lime Mortar Improved by Sulphates.— It will be seen that with $4\frac{1}{2}$ per cent. of sulphate, even with double the volume of sand, the mortar was much stronger than when lime alone was employed. It is to be regretted that no results are given with other proportions of sulphate. This mortar was found not to be adapted for use under water. The explanation of the foregoing results is surrounded with many difficulties, as the chemical action is quite different from that of selenitic cement, in which the slaking action is controlled by this means. We must probably seek for the solution in connection with the silica and alumina compounds of the Halkin lime, and it would be interesting to ascertain the behaviour of other hydraulic limes treated in a similar way.

EXPERIMENTS WITH HALKIN LIME MORTAR WITH AND WITHOUT
SULPHATE OF LIME.

DESCRIPTION OF MORTAR.	Lime.	Sand.	Ashes.	Mixed with Salt or Fresh Water.	Number of Lbs. required to Break by Tension Briquettes of 2¼ Square Inches.*			
					21 Days	42 Days	84 Days	168 Days
Ordinary mortar, ...	1	2	½	Fresh.	60	130	153	248
Do. do., ...	1	2		Salt.	47	74	134	147
Do. with 4½ per cent. plaster,	1	2		Fresh.	98	141	340	375
Do. do.,	1	2	No ashes used with the sulphate.	Salt.	67	183	232	387
Do. do.,	1	3		Fresh.	122	156	306	398
Do. do.,	1	3		Salt.	140	210	376	...
Do. do.,	1	4		Fresh.	97	154	299	422
Do. do.,	1	4		Salt.	100	190	401	...
Do. do.,	1	5		Fresh.	65	193	253	360
Do. do.,	1	5		Salt.	70	232	312	320

Cements from Sewage Sludge.—While treating of General
Scott's cement, we must briefly allude to his proposals for the
manufacture of cement from sewage sludge, a plan which was
actually carried out upon a considerable scale at Birmingham,
Burnley, Ealing, and several other towns. This invention was
a development of the so-called "lime process" of sewage treatment. In accordance with this process, a small quantity of lime
(from 15 to 23 grains per gallon) in a fine state of subdivision is
added to the sewage water, and as this substance becomes mixed
with the water in the sewer it gives rise to a copious precipitate,
consisting mainly of carbonate of lime, together with a little
phosphate of lime. The sewage is then allowed to pass into
tanks where the flow is arrested, and the precipitate slowly sinks
to the bottom, and in so doing it entangles and carries down
with it all the suspended impurities, leaving the supernatant
water thoroughly clarified, and in a fit condition to be discharged
into rivers of large volume. The sewage sludge or mud which
remains at the bottom of the tank, when this clear effluent has
been drawn off, is dried either on hot floors or by mechanical
pressure, or other suitable means, and when sufficiently dried it
is burnt in down-draught kilns of special construction, with
small quantities of interstratified fuel.

Character of Sewage Cement.—General Scott was able to
produce in this way a cement which had approximately the
composition of Portland cement, because the detritus and clayey
matters washed from the surface of the roads, and the ash of the

* In all cases the average of 3 tests.

fæcal, and other organic substances always present in sewage
water furnished the amount of silica, alumina, and iron needed
for the cement.   In dry weather it became necessary in certain
cases to add a small quantity of clay to the sewage water along
with the lime, this addition of clay aids in the clarification of
the sewage, and need rarely exceed a few grains in the gallon.
This plan of dealing with sewage had manifest advantages, as it
enabled municipal authorities to get rid of the sludge without
expense, and lime is well known to be the cheapest and best
precipitating agent.   It was found in practice that from 1 to 1½
tons of quicklime were needed for the treatment of one million
gallons of sewage water, or say for the daily volume of sewage
produced by a population of 25,000 persons.   If the cost of lime
be taken at 15s. per ton, this implies a cost of 9d. per diem per
thousand inhabitants.   From each ton of lime used with sewage
of average composition there will result 30 tons of wet sludge or
3 tons of dry sludge.   This sludge when calcined will yield about
1½ tons of cement.

Quality of **Cement from** Sewage **Sludge.**—The quality
of the cement made in this way will manifestly greatly depend
upon the composition of the sewage water, and the impurities
present therein.   By carrying out the calcination at a lower
temperature, the sludge may be converted into hydraulic lime,
or a lime suitable for agricultural use.   It was found by experiment that, for each ton of lime introduced into the sewers, 45
bushels of good agricultural lime, valued by Dr. Voelcker at 1s.
per bushel, were obtained.   The cement made at Burnley which
was calcined at a high temperature, though it fluctuated somewhat widely in composition, owing to the varying quantities of
detritus carried into the sewers during rainy weather, commanded
a ready sale, and showed a fair tensile strength.

Cement from Alkali Waste.—By means of certain patented
processes Messrs. Chance are now enabled to extract from the
alkali waste, resulting from the manufacture of soda on the
Leblanc system, a very large part of the combined sulphur which
was present as sulphate and polysulphides of calcium.   This
discovery may hereafter provide a mode of utilising these vast
mounds of bye-products which now encumber the ground, and
so sadly disfigure the vicinity of alkali works.   In order to find
some means of using the lime after the sulphur has been extracted, it has been more than once proposed to employ this
material in the manufacture of Portland cement, but the percentage of sulphur still remaining in the waste has hitherto
rendered this out of the question.   Recently, however, by a modification of the treatment, Mr. C. Spackman, F.C.S., appears to have
been successful in producing Portland cement from this material,

and from a paper published by him in the *Journal of the Society of Chemical Industry* of June 30th, 1892, we have extracted the results of certain experiments made by him with a sample of alkali waste which had the following composition :—

Coke,	2·492
Sand,	1·094
Silica,	1·156
Alumina,	0·926
Ferrous sulphate,	1·488
,, sulphide,	0·421
Calcium carbonate,	68·861
,, sulphate,	4·735
Magnesium carbonate,	2·428
Soda,	0·962
Water,	15·714
	100·277

For reasons given below this material after being dried and ground was mixed with a marly clay, containing 30 to 40 per cent. of carbonate of lime, and a considerable excess of water, which was allowed to drain off; advantage being thus taken of the slight solubility of calcium sulphide. The dried slurry thus produced was found after careful calcination to yield a cement which stood high tests and gave good results when mixed with sand. As this substance proved very fusible, considerable care was needed in the burning and the mixture with a clay rich in carbonate of lime was rendered necessary to bring down the proportion of calcium sulphate in the resultant cement.

Composition of the Cement.—The dried slurry was found to have the composition shown in Col. 1, and in Col. 2 we give Mr. Spackman's analysis of the cement :—

	No. 1.	No. 2.
Coke, water, and organic matter,	1·032	...
Sand,	1·349	2·749
Silica,	12·220	18·215
Ferrous oxide,	3·427	6·563
Ferrous sulphide,	0·324	...
Alumina,	4·993	8·048
Calcium carbonate,	71·002	...
,, sulphate,	2·455	5·006
Lime,	...	56·682
Magnesium carbonate,	1·955	...
Magnesia,	...	1·639
Potash,	0·671	0·654
Soda, &c.,	0·772	0·779
	100·200	100·335

Mr. Spackman states that if it is so far free from sulphur compounds as to give in the cement a quantity not exceeding 5 per cent. of calcium sulphate, alkali waste may, after treatment by the Chance process for the recovery of its sulphur, be successfully utilised for the manufacture of Portland cement. We learn, indeed, that Messrs. Chance have themselves recently established the manufacture of cement from their waste, upon a commercial scale, and that they are now producing large quantities of Portland cement of excellent quality from this material.

CHAPTER XV.

THE PLASTER CEMENTS.

Plaster of Paris as a Cement.—A whole series of cements, which have as their sole or chief ingredient sulphate of lime or plaster of Paris, depend for their set or induration upon entirely different properties to those which influence this reaction in the case of cements prepared from carbonate of lime. Calcic sulphate is found in a natural state in a great variety of forms, and is a substance of very common occurrence. As gypsum, the hydrated sulphate, the state in which it is perhaps most widely distributed, it contains in each 100 parts 32·60 of lime, 46·50 of sulphuric acid, and 20·90 parts of water, the chemical formula being $CaOSO_3 + 2H_2O$. There are numerous varieties of gypsum, some of them being transparent and crystalline as the selenite, some amorphous, and some fine grained and compact as in the alabaster and plaster stone.

Calcic sulphate also occurs more rarely in the anhydrous state, devoid of water, when it is known as anhydrite, and consists of a crystalline mass with a regular cleavage into rectangular prisms. This mineral has not, so far as we know, been employed for industrial purposes.

Properties of Calcic Sulphate.—Sulphate of lime has a specific gravity of 2·31. It is slightly soluble in water (much more so than the carbonate). At a temperature of 60° F., 1 part of sulphate of lime will dissolve in about 490 parts of water. As the temperature rises the solubility increases until the water reaches 100·4° F., when the solubility again decreases. At the boiling point 1 part of sulphate will be contained in 571 parts of water, but by long-continued exposure to warm water the proportions may be varied somewhat considerably, and sulphate freshly precipitated is more soluble than gypsum. The raw gypsum or plaster stone is less soluble, moreover, than that which has been calcined. The industrial value of the sulphate of lime consists in the fact that the 2 equivalents of water with which we have seen that it is combined may be expelled at a temperature of about 392° F. A large part of this water may be driven off at a very much lower temperature, and where the material is reduced to powder and kept well stirred about three-fourths of the water may be expelled at a little below the boiling point of water, or about 205° F.

M. Le Châtelier's Experiments.—M. Le Châtelier, a French savant, who has recently carried out some interesting experiments in the dehydration of the plaster stone, and whose observations were communicated to the Académie des Sciences in 1883, has remarked that there are two distinct periods of rest during the process of expelling the water. His mode of demonstrating this fact was as follows :—Some pulverised gypsum was introduced into a hard glass tube, immersed in a paraffin bath, and gradually heated, the temperature being read off at regular intervals of time, by means of a thermometer embedded in the plaster. By constructing a curve, showing graphically the increase of temperature for regular periods of time, it was seen that the heat rose rapidly to 230° F., then more steadily from 230° to 248°, remained for some time nearly stationary between 248° and 266°, then rapidly increased between 266° and 284°. A second, but minor, halt in the process took place between 320° and 338°. These two interruptions indicate the absorption of heat which accompanies the elimination of water, and point to the existence of two different hydrates, whose decomposition takes place at differing ranges of temperature.

The first of these compounds is represented by the formula commonly used for gypsum—viz., a hydrate in which 1 part of the calcic sulphate is combined with 2 parts of water $CaSO_4 + 2H_2O$. This substance parts with three-fourths of its water to form the second hydrate, in which 2 parts of the sulphate are united to 1 equivalent of water, which compound would be represented by the formula $2(CaSO_4) + H_2O$.

Commercial Plaster is a Hydrate.—The existence of this hydrate was clearly shown by M. Le Châtelier, for on heating for some time 10 grammes of powdered gypsum at a temperature of 311° F., which is intermediate between that needed for the decomposition of the two forms of hydrate, he ascertained that the loss of weight was uniformly 1·56 grammes, which corresponds very precisely with $1\frac{1}{2}$ equivalents of water, and from this it is clear that the compound thus obtained contains only half an equivalent of water, combined with the sulphate, or about 6·2 per cent. Ordinary commercial plaster of Paris as prepared for use contains, as a rule, about 7 per cent. of water, and hence it consists almost exclusively of this particular hydrate.

Plaster Burnt at High Temperatures.—On submitting the plaster to a temperature of 338° F. and upwards no change is apparent, so long as the heat does not exceed 390° to 430°, but beyond this point the material begins to lose certain of its essential properties. It ceases to absorb water with avidity, and sets only after a long interval of time. If the heat reaches 650° it cannot be gauged in the ordinary way, and comports itself exactly like anhydrite. In this state it is termed "dead-burned."

Overburnt Plaster Combines very slowly with Water.— It has sometimes been asserted that plaster thus treated can no longer combine with water, but this is an error, as the hydration can still take place if the substance bo reduced to a very fine powder; the process is, however, greatly retarded. On continuing the calcination of the plaster to the temperature of bright redness, the substance melts into a vitreous paste, which forms a crystalline mass on cooling, having all the attributes of anhydrite, which is no doubt a native gypsum, acted upon by volcanic or similar agencies. It is impossible to decompose this substance by heat, but if strongly heated in contact with charcoal, or in the presence of decomposed organic matters, it loses part of its oxygen and is converted into calcium sulphide, which substance is in turn acted upon by carbonic acid and water, giving rise to the evolution of sulphuretted hydrogen gas. It is in this way that we are able to explain the presence of sulphuretted hydrogen gas in certain mineral waters, originally containing sulphate of lime in solution. When waters rich in dissolved sulphate of lime are used in boilers, or when sea water is employed, a deposit or scale is formed, which consists mainly of the hydrated sulphate with half an equivalent of water, or one in which water is present, to the extent of some 6 or 7 per cent. The mean of several analyses of boiler-scale from a marine boiler show this substance to have the following percentage composition :—

Carbonate of lime,	.	.	.	.	.	.	0·3
Peroxide of iron,	.	.	.	.	.	.	2·0
Water,	.	.	.	.	.	.	5·8
Sulphate of lime,	.	.	.	.	.	.	91·9
							100·0

Preparation of Plaster requires great nicety.— It will be evident from the foregoing observations that the preparation of plaster of Paris is a matter of considerable nicety, and that the degree of calcination requires attention and care, for while, on the one hand, too high a temperature produces a more or less inert substance; the failure to expel the water of hydration renders the material useless for moulding and casting, which, at any rate in this country, forms one of its chief uses. In many parts of France coarse plaster takes the place of lime mortar, and impure sulphates, or those mixed with small quantities of slaked lime, appear to be well adapted for this purpose. Moreover, precautions must be taken in burning the plaster to avoid direct contact with the fire, as carelessness in this respect leads, as we have seen, to the formation of sulphide of calcium ; a very objectionable product.

Preparation of Plaster formerly practised.—The oldest and simplest plan of burning plaster-stone, and the one which is still most commonly employed abroad is effected in a rude shed,

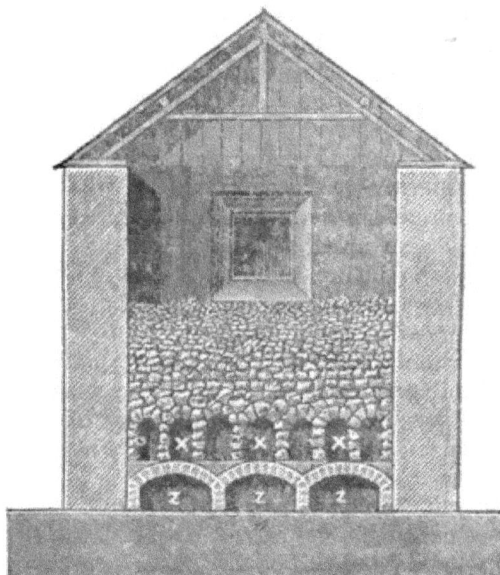

Fig. 25.—Covered Hovel for Burning Gypsum.

enclosed on three sides by walls of brick or stone, and roofed in to protect the contents from the weather.   As will be seen by our illustration, the floor is sometimes hollow, and upon it the stone is built in a series of rough arches, x,x,x, disposed parallel to the main walls of the building, these arched flues are constructed of the largest lumps of stone, and immediately over them are placed the lumps next in size, the smaller pieces being next selected, and the size of the pieces of stone decreasing upwards as the heat from the fire diminishes, the top of all being covered in with dust and sweepings.   Wood fires are then lighted in each flue, and these are regulated so as to produce a moderate heat, and kept going as steadily as possible in order to maintain an equable temperature throughout the entire mass of stone. The flames pass upwards through the interstices of the material and drive off the moisture and the water of hydration, which escape in the form of dense vapours, and pass out through openings in the roof and in the upper part of the shed.

Plaster Burnt with Coal.—When coal is employed in the calcination, rude furnaces are formed in the arched spaces beneath the floor, as seen at z, z, z in our illustration, and the flames pass through special apertures contrived in the tops of the arches. The firing is continued until the arch stones at the base of the mass show visible redness, say for about twelve hours, after which the fires are drawn and the plaster is allowed to cool slowly. It will readily be seen that this plan is a very unscientific one, for while the top layers are often barely deprived of their combined water, the lumps forming the voussoirs of the arches are overburnt, and yield a partially if not wholly inert plaster. In spite of every care also, it is impossible to prevent the formation of a certain percentage of calcium sulphide, which gives rise to the unpleasant smell of sulphuretted hydrogen always apparent when plaster made in this way is gauged with water, and the presence of this sulphide is very unfavourable to the setting of the compound.

Preparation of Plaster in the Vicinity of Paris.—A much better system of dehydration is that practised in the vicinity of Paris, for which a kiln is employed somewhat resembling in form those used for lime, but furnished with an arched cavity at the base, formed of firebricks and pierced with openings for the passage of the flame and heat. In a regular kiln of this description the heat can be more evenly distributed, but it has the same objection as the one we have previously noticed, in that the bottom layers receive much more heat than those at the top, and the product, even with the utmost care, is never uniform throughout.

Plaster Kiln of Improved Construction.—A kiln which is much better in principle, in that the heat is under more complete control, and can be imparted more evenly to the whole of the contents, is that invented by M. Dumesnil, shown in Figs. 26 and 27. This kiln is circular in plan and has a central furnace, D, above which is the fire chamber, G, formed of fire-brick and furnished with twelve openings at F, F, F; the flame passes from the furnace into the chamber, G, by curved flues, seen in section at E, E. Each of the openings at F is connected with a radiating flue, M, constructed with lumps of plaster built into the form of an arch. Above these flues the stone is arranged in layers, R, S, T, the larger pieces of stone being placed near the bottom and the smaller fragments in the upper layers. The top of the kiln is arched, and has one central and four smaller flues, all of which can be closed by means of dampers. In the arched roof, L, L, is an orifice, N, for filling the top of the kiln, while a door

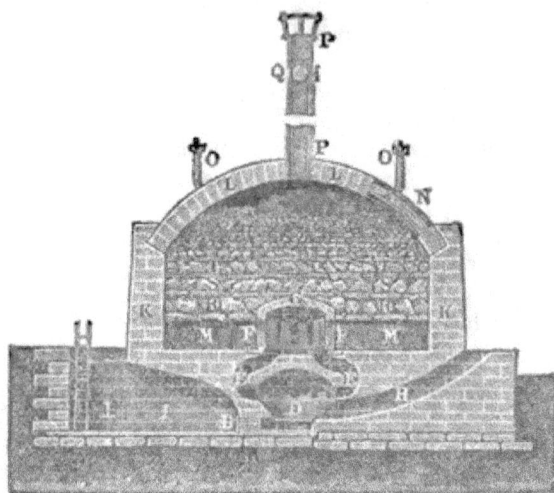

Fig. 26.—Transverse Section of the Dumesnil Kiln.

in one side serves for emptying and loading. The stoke-hole is shown at H, and the ashpit at I. In burning this kiln, which is 20 feet in diameter and 13 feet to the top of the arch, the

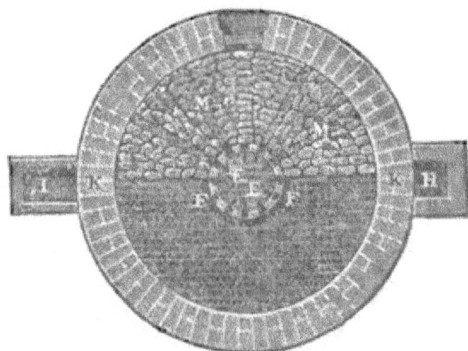

Fig. 27.—Plan of Plaster Kiln.

contents are dehydrated by means of a moderate fire in about twelve hours, and then the fires are extinguished and an extra charge of 6 or 7 cubic yards of stone is added above the top

layers, after which all the openings are closed, and the heat of
the charge is sufficient to effect the dehydration of the topping.
The contents of this kiln are found to be about 45 cubic yards.
The kiln is economical to work in the matter of fuel-consump-
tion, and the products are fairly uniform in quality, but the first
cost of its erection is very considerable.

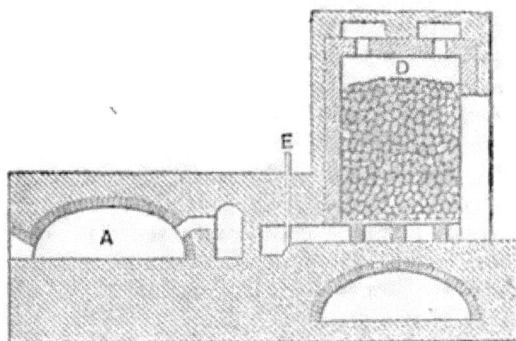

Fig. 28.—Section through Combined Coke Ovens and Kilns.

Other methods of Burning **Plaster Stone.**—Various other
plans for burning plaster stone have been from time to time
introduced, some of them with a fair measure of success. Thus
the waste gases from coke ovens have been utilised to heat the

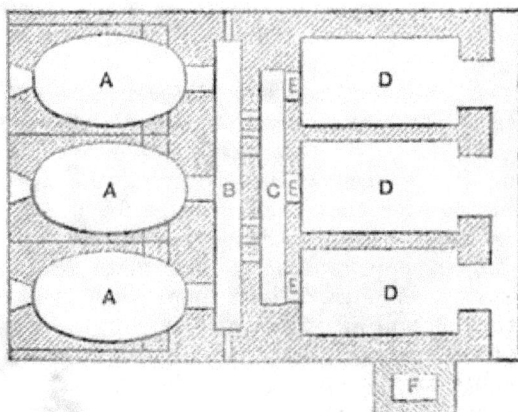

Fig. 29.—Plan showing Combined Coke Ovens and Plaster Kilns.

stone, and the arrangement of the kilns and ovens for this
purpose is shown by Figs. 28 and 29. Here A, A, A are coke
ovens of an ordinary type which discharge their surplus heat

into a collecting flue, B. This flue communicates with a subsidiary flue, C, leading to the plaster kilns, D, D, D. It is possible, by means of dampers at E, E. E, to turn the heat into any one or more of a series of kilns. The floors of the kilns are perforated, and the flames pass up into the charge of plaster stone, which is piled up to a considerable height. The steam and waste heat finally escape by means of a collecting flue running along the tops of the kilns into the chimney, F.

Use of Superheated Steam.—Superheated steam and gas have also been employed successfully. When steam is used it is raised to a temperature of about 390° F. and blown alternately into each of two chambers filled with raw stone. The high temperature of the steam rapidly abstracts the water of hydration and leaves the plaster in a condition ready for grinding. Some experiments upon a small scale conducted by Mons. Violette led him to the conclusion that 3 cwts. of gypsum could be dehydrated in three hours by a current of superheated steam, amounting in weight to about 132 lbs.

Several processes have been brought forward for dealing with the plaster stone in continuous kilns, or " running kilns," resembling somewhat those used by lime burners, and a process of roasting in cylinders caused to rotate, or furnished with a creeper or spiral screw, so as to move the plaster continuously forward from one end to the other has also been tried. Many of these plans have resulted in the production of plaster of good and uniform quality, and free from the impurities and imperfections arising from the crude and unskilled processes in common use.

Best Plaster prepared by first Grinding the Raw Stone. —For plaster of the best description, capable of setting quickly, and ensuring hard and durable casts, moulders prefer to obtain the raw stone for themselves, and to grind the same to a fine powder. This powder they then prepare for use by a so-called process of " boiling." The plaster meal is spread in a layer, some 2 or 3 inches in depth, upon a hot plate or in a shallow metal dish over a fire ; in a short time, when the temperature approaches that of boiling water, a strange motion is communicated to the whole mass of the material, and the surface appears to rise up bodily as if suspended by the aqueous vapour given off by the lower layers. Little openings or craters are formed all over the surface, and the steam passes off freely, mingled with fine dust. From time to time the plaster is stirred, and care is taken to avoid an excess of heat. When no further evolution of moisture takes place, which can be tested by holding over the surface a cold plate of glass or metal, to condense the steam, the heat is withdrawn, and the plaster is ready for use. When prepared in this way the combination with the water used for gauging takes place with great rapidity, and sound and hard casts are produced.

Plaster Baked in Special Ovens.—Much of the plaster used in this country is baked in ovens, constructed on the principle of the baker's oven. In this case the stone is broken into lumps about the size of a small hen's egg, and introduced into a well-heated oven, the temperature being but little greater than that used in baking bread. It may be tested by inserting the hand, which should be able to bear the heat for two or three seconds. The oven is then closed up, and specimens are from time to time withdrawn, to ascertain if the dehydration is complete, the baking lasting from twelve to twenty hours. The man in charge of the operation is able to judge, by the colour and appearance of the lump removed, if the water is expelled. Properly burnt stone exhibits on the white earthy fractured surface only a few bright specks here and there of crystalline particles, not completely deprived of the water of hydration. When the stone has been cooled it is ready for grinding.

The Grinding of Plaster for Use.—The grinding of plaster is a very simple matter, and almost any contrivance for this purpose will give good results. In some cases edge-runners or rollers are used, but the best meal is produced by mill-stones, such as are employed for grinding flour. Evenness and regularity of grain are most essential in the case of the best quality of plaster used for casting purposes. Such plaster has a soft, smooth feel to the touch, and sticks slightly to the fingers. It has, in fact, rather a tendency to cake together when compressed, while plaster which has not been sufficiently dehydrated has a dry and slightly gritty feeling. Overburnt plaster shows little tendency to absorb water, and cracks when made up with water. The ground plaster should be kept as far as possible from contact with the air, and not merely in sacks, as is too often the case. The powder has a strong tendency to absorb moisture, after which it becomes slow-setting, and yields a less solid casting. In France heaps of plaster are sometimes kept in good condition for a long period (even for a twelvemonth) by slightly wetting the outside of the mass with a watering can, by this means a preservative crust is formed which suffices to protect the interior of the heap.

The Gauging of Plaster.—The gauging of plaster, as the mixing of the same with water is termed, always gives rise to an elevation of temperature, due to the hydration of the calcic sulphate, and is a similar reaction to that which takes place when lime is slaked. We have seen that, in accordance with the tem-

13

perature at which it has been burned, the gypsum is more or less fully deprived of its water of hydration, and that its recombination with water takes place with so much energy as to cause a considerable rise in the temperature of the mass. There is, at the same time, an increase in bulk, which may amount to as much as 1 per cent. in 24 hours after gauging. The volume of water employed has some influence upon the rapidity of the set, and upon the ultimate hardness of the plaster, but even with a very considerable excess of water the plaster has still the power of setting, which we know is not the case with quicklime. The commonly received opinion respecting the setting of plaster is that the process is not only one of hydration, but that simultaneously a crystallisation of the mass takes place. The particles of the powder are converted into a porous network of crystals, which enclose in their interstices a certain proportion of the water containing sulphate of lime in solution. This water is held mechanically, and speedily evaporates, causing the dissolved sulphate to crystallise out, and thus adds to the hardness of the already-formed mass.

M. Landrin's Investigations into the Set of Plaster.— M. Landrin, who has investigated the behaviour of plaster with water microscopically, confirms this opinion, and assigns three distinct phases to the operation of setting. First, the plaster assumes on the contact with water a crystalline structure; second, the surrounding water dissolves a portion of the sulphate of lime; and thirdly, a part of the liquid evaporates owing to the rise in temperature caused by this chemical action, a crystal is formed and determines the crystallisation of the entire mass, in consequence of a phenomenon analogous to that which is observable when a crystal of sulphate of soda, containing 10 molecules of water, is thrown into a saturated solution of that salt.

M. Le Châtelier's Researches.—It appears that in 1883 Mons. Le Châtelier, in a communication upon this subject to the Académie des Sciences, pointed out that a direct transformation, according to the theory hitherto accepted, of the solid anhydrous sulphate of lime into the solid crystalline hydrated sulphate, would constitute an exception to the general law of crystallisation, and, moreover, the mere fact of crystallisation would not necessarily entail the aggregation of the mass. Thus, for instance, sulphate of lime, precipitated by means of alcohol from a concentrated solution, presents a maximum amount of entanglement of the crystals, but the precipitate when dried shows no tendency to cohere together into a solid mass.

New Theory as to the Process of Setting.—In order, therefore, to account for the setting of plaster and of other analogous substances, M. Le Châtelier has been compelled to formulate a new theory, based upon the phenomena of supersaturation investigated by **Marignac.**

Observations by M. Marignac.—This observer **has shown** that the hydrated calcic sulphate with half an equivalent of water, which remains undecomposed at a temperature of about 310° F., dissolves freely when shaken up with water, but that after a short interval the solution becomes turbid. This is due to the formation of a crystalline precipitate of the common hydrate with two equivalents of water, which has the formula of gypsum. The solution formed in the first case is five times as concentrated as that made from the less completely hydrated sulphate. It would appear from this that the most important agent in the accomplishment of the setting process is the relatively soluble hydrate—namely, that with a small percentage of water. This hydrate is at once dissolved, and then gives rise to the formation of the hydrate with the full equivalent of water. This latter compound decreases the solubility of the mixture, and the water becomes supersaturated with the $CaSO_4 2H_2O$ hydrate, which crystallises out. This process continues so long as there remains any of the soluble hydrate $[2(CaSO_4)H_2O]$ to fortify the solution.

The set of plaster is thus the result of two distinct series of operations, which take place simultaneously; first the particles of sulphate of lime in the act of hydration are dissolved in the water used to gauge them and produce a supersaturated solution; the solution thus formed deposits crystals of the hydrated sulphate. These crystals gradually increase in size, and form a compact mass, in the same way as all similar crystals deposited slowly from a saline solution, and this process is continued as long as any of the more anhydrous sulphate remains available to become dissolved and to keep the solution supersaturated.

Connection between this Process and the Induration of Cements.—This theory has, we believe, a somewhat important bearing on the set of calcic silicates, and the transference of the soluble lime to the crystalline silicates may take place in a somewhat similar way. It is on this account that we have reproduced, at some length, the valuable remarks of Mons. Duquesnay in the *Encyclopédie Chimique* treating of this subject.

Substances used to improve Plaster.—It has long been known that a variety of substances are capable of imparting to the somewhat soft and friable composition of the set plaster a greatly increased hardness and consequent durability. Gay-Lussac has pointed out that the hardest crude gypsum yields, after calcination and reduction to a powder, the hardest casts. This same property, so far as mortar was concerned, was, as we have seen, erroneously ascribed to dense and hard limestones. We learn from Tissot that if burnt gypsum, after it has set, be repeatedly steeped in water and allowed to dry between each soaking, it will greatly improve the crystallisation of the mass and result in a much harder casting.

Alum used with Plaster.—The most common method, however, of hardening plaster depends upon the employment with it of alum, under a plan proposed, in the first instance, by Pauware, and improved upon subsequently by Greenwood. Alum may be used in two ways, either the finished casting may be steeped in a strong solution of alum, and then slowly dried in a current of warm air—a process which needs at least a month in the bath—or the lumps of plaster when withdrawn from the kiln may be treated with a solution of alum, and then again raised to a red heat in a suitable kiln or oven. The heat in this second burning must be much greater than that needed for the dehydration of the gypsum. It is very necessary also that the heat during this second firing should be steady and uniform. The lumps when properly burnt have a dull milk-white or even a pale yellowish tint, but if the calcination is carried too far the lumps become as hard as stone, and are very difficult to reduce to powder. A composition of gypsum and alum, burnt at the requisite temperature, is readily ground, and when pulverised and gauged with water sets as rapidly as common plaster, but the resulting cement is not remarkably hard, unless the water employed consists of a solution of alum, containing from $\frac{1}{12}$ to $\frac{1}{13}$ by weight of alum. The casts obtained by this process continue to give off moisture or to "sweat" for a longer period than those made with ordinary plaster, but they acquire in time a degree of hardness comparable with that of alabaster or even marble, and they are capable of receiving a high polish. The surface has a creamy tint, and objects cast in this alum plaster, being much less soluble, will bear exposure to the weather, and will even resist the prolonged action of boiling water. It has been observed that these castings are semi-transparent, and transmit a certain amount of light through the thinner portions.

The Theory of the Action of Alum.—Various reasons have been alleged for this action of alum on plaster. Payen supposed that the induration was owing to the formation of a double sulphate of lime and potash, the crystals of which were embedded in a precipitate of alumina, but this surmise has been controverted by the investigations of M. Landrin, who found by the analysis of numerous specimens of plaster treated with alum, both of French and English manufacture, that these substances were almost absolutely pure and free from alumina and potash. The results he obtained are seen in the accompanying table. Nos. 1 and 4 are French, and Nos. 2 and 3 are of English make.

Alum Cement.	Sulphate of Lime.	Carbonate of Lime.	Silica.	Water.	Total.
1, . . .	96·75	1·05	0·72	1·48	100
2, . . .	98·19	0·41	...	1·40	100
3, . . .	98·02	0·37	0·42	1·19	100
4, . . .	98·05	0·36	0·51	1·08	100

New Theory by M. Landrin.—Seeing that these samples were all so free from water, it may be assumed that they were in each case burned at a high temperature, and the absence of potash and alumina led M. Landrin to propound an entirely different hypothesis. He came to the conclusion that the action of the alum was caused by its contents in sulphuric acid, rather than by the bases present therein, and that this acid led to the conversion of the carbonate of lime into a sulphate of lime. In order to ascertain the accuracy of this surmise, he made use, in the first place, of a number of soluble sulphates, such as the sulphates of soda, potash, and ammonia, and caused them to act upon common plaster. He took care to employ only the precise amount of each of these substances which would furnish the supply of sulphuric acid needed to act upon the carbonate of lime, and he obtained precisely the same results as with alum. Parallel experiments, with sulphuric acid alone, gave, as he anticipated, corresponding results, and he was thus enabled to indicate a new process for the preparation of alum-plasters. All that is needed to impart the same degree of hardness as that due to the use of alum is to steep the raw plaster stone for about a quarter of an hour in a 10 per cent. solution of sulphuric acid and then to fire it at a dull red heat. By this means he obtained an excellent plaster cement, which left nothing to be desired in point of hardness, and which was sufficiently slow-setting. Moreover, the effect of the dilute acid was to destroy all traces of organic matter (always found in the raw gypsum, and which tend to give it a greyish colour), and to bleach it most perfectly, so that the resulting cement was exceptionally white and pure.

Importance of Expelling the Acid.—It must be remembered, in connection with this process, that it is absolutely essential that all the uncombined acid should be expelled, as even slight traces of sulphuric acid would render the plaster more or less hygrometric and liable to attract water. The burning should, on this account, be carried out at a temperature of from 1080° to 1260° Fahrenheit. As the effect of adding the alum is, as we have thus seen, merely to augment the amount of sulphate of lime present in the plaster, we can only attribute the slowness of the set to the influence of the high temperature at which it is burned, and there is thus a complete uniformity of action between the alum or sulphate plaster and that prepared in the ordinary way.

Keene's Cement.—Several other plaster cements have become widely known in this country, and they are for the most part prepared in a similar way to that we have described in the case of M. Landrin's process, only using a different solution. Keene's cement is made by steeping the calcined stone in a strong solution of borax and cream of tartar. The liquor is composed of 1 part of borax and 1 part of cream of tartar, dissolved in about 18 parts of water. In this solution the plaster in the lump, as withdrawn from the oven, is allowed to remain until it is thoroughly impregnated with the salts. It is then taken out, dried, and reburned at a temperature of dull redness for about six or eight hours. When cool it is ground to a fine powder, and it is ready for use. It is found that borax alone gives equally good results, and the more concentrated is the solution into which the plaster is introduced the slower is the ultimate set. Thus, if to 1 part of a saturated solution of borax we add 12 parts of water, and employ this liquid as the bath, the set will take place in about fifteen minutes; but if only 8 parts of water be used, the cement will take at least an hour to set; and if 4 volumes of water be used, the cement will only become set after the expiration of several hours. The manufacture of Keating's cement is similar in all respect to the process employed by Keene.

Martin's Cement.—In the preparation of Martin's cement the solution employed is one of carbonate of potash, the stages of the manufacture being similar to those already described. For Parian cement, the bath employed may contain borax, but we understand that it is also prepared by calcining an intimate mixture of powdered gypsum and dry borax, which materials are subsequently ground to a fine powder, which constitutes the finished cement.

Respecting the chemistry involved in the production of these cements, Knapp remarks that their action is probably due to the fact that one equivalent of water contained in the gypsum is capable of being replaced by a saline compound, and that possibly these substances have the power of completely taking the place of a portion of the water, and thus give rise to the greater degree of hardness attained by the compound. A solution of tartrate of potash and soda (seignette salt) causes plaster to set instantaneously. The indurated mass has the appearance of ordinary gypsum; but it possesses the property, when it is repulverised, of again becoming hard when moistened with a solution of a salt of potash. Knapp's theory, it may be remarked, does not agree with that of M. Landrin. Many of these reactions are very interesting and merit further study, which might lead to important industrial results.

CHAPTER XVI.

SPECIFICATIONS FOR PORTLAND CEMENT.

Differences Prevalent in Specifications.—There are few subjects concerning which English engineers are more at variance than cement specifications. In the course of a discussion upon this question at the Institution of Civil Engineers in 1880, Mr. William Gostling, the well-known manufacturer, gave the following examples of diversities of opinion of various engineers, as shown in the three cardinal tests of fineness, weight, and tensile strength. As will be seen, he analysed for this purpose twenty-one specifications :—

1. *Fineness.*

In 3 specifications, no mention of fineness.

5	,,	finely ground.
1	,,	very finely ground.
2	,,	extremely finely ground.
2	,,	residue not to exceed 10 per cent. with sieve of 1,600 meshes per square inch.
1	,,	residue not to exceed 20 per cent. with sieve of 2,500 meshes per square inch.
1	,,	residue not to exceed 15 per cent. with sieve of 2,500 meshes per square inch.
1	,,	residue not to exceed 10 per cent. with sieve of 2,500 meshes per square inch.
1	,,	90 per cent. to pass through a sieve having apertures of not more than $\frac{1}{18}$ inch in diameter.
1	,,	no residue on passing through a sieve of 50 meshes to lineal inch.
1	,,	80 per cent. to pass through seive 5,800 meshes per square inch.
1	,,	85 per cent. to pass through sieve 5,800 meshes per square inch.
1	,,	fineness to be satisfactory to inspector; unsatisfactory if tests made from screened cement are stronger than those made from unscreened.

21 specifications, with thirteen varieties of test.

Differences as respects Fine Grinding.—It was scarcely necessary, he said, to point out the inconvenience of requiring fine grinding merely in general terms. Such specifications betrayed an inadequate estimate of the value of fine grinding, and opened a wide field for disputes. Of the other specifications, two indicated the use of sieves of forty wires per lineal inch, and three required sieves of fifty wires. In using these the percentages varied, and only one specification demanded complete sifting. An important specification, lately issued, assumed either that there was no appreciable difference between screened and unscreened cement, if finely ground, or that, if there was a difference, unscreened cement was the stronger; assumptions which were open to question. Then there was, as we see, an attempt to introduce the use of sieves of 5,800 meshes to the square inch, hitherto unused by the trade in England. All recent investigations in Germany and England demonstrate the importance of fine grinding, and surely it is time that this fact should be recognised, and that a uniform system, reconciling practice and theory, should be established and applied in drawing specifications.

2. *Weight per bushel.*

In 1 specification not less than 100 lbs. per striked imperial bushel.

1	,,	,,	110 ,,	,,	,,
8	,,	,,	112 ,,	,,	,,
1	,,	,,	114 ,,	,,	,,
4	,,	,,	115 ,,	,,	,,
1	,,	,,	116 ,,	,,	,,
1	,,	about	118 ,,	,,	
1	,,	not less than 108 ,,	nor more than 115 lbs. per imp. bush.		
1	,,	,,	112 ,,	,,	118 ,, ,,
2	,,	without any reference to weight.			

21 specifications, containing ten varieties of tests.

Variations in the **Weight Test.**—Sixteen of these specifications placed no limit to excess of weight per bushel, merely guarding against undue lightness. One specification demanded a specified weight, and two specifications only adopted the more rational method of confining the weight within definite limits. There was no guiding principle apparent. The fact that the weight should bear a definite relation to the strength and fineness was not recognised, and weight of itself appeared to be deemed an indication of quality. The new specifications of one important public body discarded the test of weight entirely, rightly deciding that fineness of grinding, and strength, were the true tests. Assuredly, if this test was to be retained, some kind of scale, by which the relation between the weight and strength might be determined, should be adopted.

3. *Tensile breaking strain, seven days after gauging.*

In 1 specification not less than 200 lbs. per sq. in., or 450 lbs. per 2¼ ins.
section.

1	,,	,,	250 ,,	,,	562¼ ,,	,,
2	,,	,,	267 ,,	,,	600 ,,	,,
5	,,	,,	300 ,,	,,	675 ,,	,,
1	,,	,,	311 ,,	,,	700 ,,	,,
1	,,	,,	333 ,,	,,	750 ,,	,,
2	,,	,,	350 ,,	,,	787½ ,,	,,
2	,,	,,	356 ,,	,,	800 ,,	,,
1	,,	,,	378 ,,	,,	850 ,,	,,
2	,,	,,	400 ,,	,,	900 ,,	,,
1	,,	,,	444 ,,	,,	1,000 ,,	,,

1 ,, cement gauged with three times its weight of sand, to stand a strain of 140 lbs. per square inch twenty-eight days after making briquettes.

1 ,, cement gauged with 3 parts sand, to bear 150 lbs. per square inch twenty-eight days after gauging, if the cement set neat in less than two hours, and 170 lbs. per square inch if taking more than two hours to set.

21 specifications, embracing thirteen varieties of tests.

Actual Points of Difference in the twenty-one Specifications. — Thus in these twenty-one specifications there were thirty-seven variations in the tests, and as a great number of specifications would doubtless add to these variations, and admit of the three tests being combined differently, it becomes difficult to estimate the number of varieties in actual use.

Everyone who has had to do with Portland cement testing thinks himself able to devise a series of tests, and in the absence of any standard specification we obtain all these wearisome and vexatious differences of opinion.

Board of Works Specification.—One of the latest specifications of the now extinct Metropolitan Board of Works read as follows :—

"The whole of the cement for these works, and herein referred to, is to be Portland cement of the best quality, ground so fine that the residue on a sieve of 5,800 meshes to the square inch (equal to about 76 per lineal inch) shall not exceed 15 per cent.* by weight. When tested, should the proportion which will not pass through the sieve be greater, a quantity of neat cement, equal to such excess, shall be added (at the contractor's expense) to the specified proportions of all cement, mortar, or concrete used upon the works. When brought upon the works it is immediately to be put into a dry shed or store, which the contractor shall provide for the purpose, having a wooden floor and all necessary subdivisions. The cement is to be emptied out upon the floor, every 10 tons being kept separate, and it is not to be used until it has been tested by samples taken from different sacks.

* London County Council specification, 10 per cent.

The cement is to be gauged with **three times its weight of dry sand**, which has been passed through a sieve of 400 and been **retained upon one of 900** meshes to the square inch. The cement and sand having been well mixed dry, about 10 per cent. of their weight of water is to be added, and briquettes formed in moulds of 1 inch sectional area at the weakest point. The briquettes, having in the meantime been kept in a damp atmosphere, are to be put into water twenty-four hours after they have been made, and remain in water until their tensile **strength** is tested by means of apparatus belonging to the Board and by their officers. These briquettes must bear, without breaking, **a weight** of 230 lbs.* per square inch, twenty-eight days after the briquettes have been made; and the cement neat must not, at any season of the year, **set in less than one hour.** Any cement which does not **answer** the requirements **will be** rejected, **and** must be forthwith removed **from** the works. Briquettes **of** neat cement shall also be made and broken **after seven** days, **as may** be directed by the engineer. Their limit of resistance and the weight **of** the cement shall also give results to his approval. The cement store to be provided **with a** Chubb's patent lock **and key, to be** obtained and fixed by the Board."

The specification issued by the London County Council is, with a few trifling particulars, to some of which we have drawn attention by foot-notes, almost identical with the above. They state, however, that—

"Besides these mechanical tests, the cement will be **tested as to its** chemical character and as to the specific gravity, **which is not to be less** than 3·10."

There is also a note in the following terms :—

"To secure **due progress** the contractor shall, immediately after the contract is signed, procure and deliver on to the site of the works a stock of cement sufficient for carrying on the works for at least five weeks, and at all times, **until their** approaching completion, **shall keep stored on the** works a quantity equal to five weeks' consumption."

We, no doubt, owe all of these specifications to Mr. Grant, and it is, therefore, advisable, while we are on this branch of the subject, to reproduce the skeleton form of specification which he appended to his paper at the Institution of Civil Engineers in 1880.

"Mr. Grant's Form of Specification for Portland Cement.

"CEMENT-MORTAR AND CONCRETE.

"*Portland Cement.*—The whole of the cement for these works, and herein referred to, is to be Portland cement of the best quality,† ground so fine

---

* L. C. C. specification, 250 lbs. per square inch is applied at the rate of 200 lbs. per minute.

† Here, if desired, the words may be inserted "weighing not less than 112 lbs. to the bushel."

that the residue on a sieve of   * meshes to the square inch shall not
be more than   † per cent. by weight.   Should the residue be greater,
a quantity of cement proportionate to such excess must be added.   When
brought upon the works it is to be put into a dry shed or store which the
contractor   to provide for the purpose, having a wooden floor and all
necessary subdivisions.   The cement is to be emptied out upon the floor,
every [   ] bushels being kept separate, and it is not to be used until it has
been tested by samples taken from different sacks.

"*Testing.*—The cement is to be gauged with three times its weight of
dry sand, which has passed through a sieve of 400, and been retained upon
one of 900 meshes to the square inch.   The cement and sand having been
well mixed dry, about 10 per cent. of their weight of water is to be added,
and briquettes formed in moulds of 1 inch sectional area at the weakest
part.   The briquettes having in the meantime been kept in a damp atmosphere are to be put into water twenty-four hours after they have been
made, and remain in water until their tensile strength is tested by means
of apparatus belonging to the Board, and by their officers.   All cement
that, when neat, sets in less than two hours, must bear without breaking,
when subjected to this test, a weight of   ‡ lbs., and if it take from two
to five hours to set neat, it must bear a breaking weight of [170] lbs. per
square inch twenty-eight days after the briquettes have been made.   If
not, it will be rejected, and must be forthwith removed from the works.§

"*Cement-mortar.*— The cement hereinbefore described, excepting as
herein otherwise specified, shall for mortar be mixed in the proportion of
1 bushel or 112 lbs.‖ of cement to [two] bushels of sand.

"*Cement Concrete.*—The whole of the concrete to be used throughout the
works is to be formed of   Imperial bushels or cwts. of 112 lbs. of
Portland cement to 1 cubic yard of ballast.   The materials to be turned
over three times in a dry state; the water then to be added, and the mass
turned over three or four times more and thoroughly intermixed."

Specification proposed by Mr. V. de Michele.—Mr. V. de
Michele, an engineer of great experience, and a manufacturer of
cement, has recently proposed the following very concise tests,
which he considers would be sufficient if embodied in a specification to "ensure Portland cement of the highest quality at
market rates:"—

Pats ½-inch thick in water, absolutely sound at 7 days.
Tensile strength, 400 lbs. per square inch at 7 days.
Fineness, 10 per cent. residue on a 50 sieve.

---

* Either 6,400 meshes to the square = 80 to the lineal inch, or 5,806 per
square inch = 76·2 to the lineal inch, which corresponds with 900 per
square centimetre.
† The residue may be either 20 per cent. as in the German rules, or 10
per cent. as in their practice.
‡ Here insert either 142 lbs. = the German minimum of 10 kilogrammes
per square centimetre, or a higher number, especially if the cement, neat,
takes more than two hours to set.
§ If a test for neat cement is inserted, the following words may be added
here—"tested neat, the cement must bear a tensile strain of at least 400
lbs. per square inch, after seven days, six of them in water; and 550 or
600 lbs. per square inch after twenty-eight days, twenty-seven of them in
water."
‖ The contractor may either measure or weigh the cement.

Note.—The pats to be gauged on glass, immersed in water immediately, and left there for the whole period. One pat to each three bricks.

The test bricks to be gauged by a skilled man, with any quantity of water, in any way he likes. The average of three to be taken, which shall represent about 100 tons or less. The strain to be applied as quickly as possible.

The sieve to have 2,500 holes per square inch, and to be of wire $\frac{1}{185}$th of an inch in diameter. Shaking to be continued until nothing ground in the mill passes.

There is no doubt that from the manufacturer's point of view such tests would be deemed sufficient, but we consider the test of fineness altogether too lenient, while the immersion of the pat in water directly it is made is only practicable in the case of quick-setting cements.

A Very Strict Specification for Cement.—The following abstract from a Specification for Portland cement to be supplied to some works in the north of England, will indicate the other extreme of rigour :—

Quality and Tests for Cements.—The cement shall be the best Portland cement, finely ground, uniform in quality, and shall be capable of bearing the following tests :—

A. Samples of the cement must pass through a sieve having 900 meshes to the square inch, and on being sifted through a 6,000 gauge wire mesh must not leave a residue of more than 25 per cent.

B. The cement must not weigh less than 110 lbs. per striked bushel.

C. Samples of pure cement will be gauged with a medium quantity of water, and placed in brass moulds. Within 24 hours of being moulded, the casts thus made will be immersed in still water, and there remain 3 days from the date of moulding, after which the tensile strength will be tested, and which must not be less than 175 lbs., and not exceed 275 lbs. per square inch of sectional area, and the same mixing tested in 7 days must stand a test increasing from 40 to 50 per cent.

D. The cement, when gauged neat in the mould, must not become firm to the finger nail in less than 3 hours after mixing.

E. The cement will be made into pats and placed in water as soon as they are set, and any cement that develops cracks, or otherwise, will be rejected.

We hesitate, with all these differences of opinion, to give our own views as to a model specification. We can only hope that unanimity may ere long be established, and that the adoption of a uniform system of testing may lead to the use of a standard form of specification.

A.

ens of Com

	TIC MORTAR. auged."	
	Test No.	St:
	G	
	715	38
Li	716	33
	718	32
	724	37
Li	725	32
	723	28
	722	32
Li	720	25
	719	14
	Mean	
P	of area	1

f Joint 18·5 sq

d apart.

we the san

Appendix B.—The Gases evolved from Cement Works.

In certain localities where cement works have been established in the midst of a residential population, the fumes evolved during the process of burning the cement and the smoke from the drying floors and chimneys have caused more or less nuisance, and have been objected to by the inhabitants. It may, therefore, be as well to consider briefly the grounds for these complaints, and to show how these evils may best be avoided or mitigated. In this enquiry we cannot do better than avail ourselves largely of the able report by Mr. A. Blaikie on *Cement Works in Kent*, which will be found in the Twentieth Annual Report on Alkali, &c., Works for the year 1883. The inspector states that "for the purpose of ascertaining the character of the kiln-gases emitted, and of tracing any changes that might take place in them from day to day as the kiln burnt out, a number of analyses has been made of washings of gases taken both from the tops of open kilns and from the flues and chambers of close kilns." He points out that "the usual time required for burning out a large kiln is from three to four days. On the first day a white vapour is given off which consists practically of nothing but moisture, almost no solid matter or smell being present. It is, however, in some states of the weather very dense. On the second day, as the heat gets up, the smell is often rank and offensive, a smell of burning organic matter is given off by the slip, and there also is some smell from the coke. This vapour is white and very heavy; it contains a quantity of solid matter and is usually slightly alkaline as the gases from the fire passing through the upper layers of dried slip are neutralised and carry away fine particles of chalk. On the third day, sometimes not till the fourth, the kiln is at its greatest heat and beginning to burn out, the vapours being very much lighter and less dense. It has much less smell than the vapour on the second day, but contains much more volatile matter, chiefly chlorides and sulphates of the alkalies, and it has been found that the quantity of chlorides in the washings always increases with the increase of solid matter. The sodium chloride is derived from the water used in preparing the slip which, as a rule, is very salt, and the potassium is probably liberated from the clay (which contains about 0·6 per cent.), by the action of the chalk. During the later stages the amount of the carbonic oxide is large and to those close by, and inhaling it, poisonous. The gases are also slightly (in most cases very slightly) acid." Mr. Blaikie shows that the nuisance is generally more acute in the case of the common open kilns than in those which are provided with a chamber on the Johnson system, or which discharge their gases into flues for drying the slurry on various principles which he describes—the plans of Messrs. Burge, White & Glover, and Margetts. In all these latter kilns there must of necessity be a lofty chimney which causes the gases to be diluted and dispersed. He traces no evil effects on vegetation arising from the fumes caused by cement works, but he says, "near one works, some years ago, there was a fine group of elm trees, and these have all been killed since the coke ovens were erected." He notices, on the other hand, the beneficial effects of these works in causing the disappearance of ague, though we think it is more than doubtful whether there is any connection between the manufacture of cement and the gradual decrease in the prevalence of this disorder. It is often stated that the vapour and fumes evolved from lime kilns have a good effect upon phthisic and consumptive patients, but we have never seen any statistics which can be quoted in support of either of these allegations.

Appendix C.—The Effects of Sea Water on Cement.

We may summarise Mr. Carey's remarks on the action of sea water as follows:—He puts the case thus, "The quantity of magnesium in sea water, present as chloride and sulphate, amounts to 0·06 per cent. by weight. In a porous mass of concrete, a precipitation of the salts of magnesia takes place, some of the lime of the cement being at the same time dissolved. On the one side, it is held that these salts of magnesia, filling the pores and interstices of the concrete, subsequently undergo a change of volume, producing disintegration. On the other hand, it is asserted that these precipitates are absolutely inert, and do not in any way affect the strength of the concrete." The views of Dr Michaelis respecting the injuries likely to arise from the presence of magnesium salts (quite from another aspect), are quoted and it is admitted that he regards the rejection of cement containing over 5 per cent. of magnesia as advisable, owing to the acids with which this body is combined. Dr. Michaelis has, however, reported that Portland cements in which "magnesium was present to the extent of 20 per cent. were under his observation for ten years and showed no signs of flaw." Special tests made to determine the action of magnesia led Dr. Michaelis to regard this substance as a species of adulterant, when it took the place of a corresponding percentage of lime. He states that it is not actively useful in cements, but that it does not lead to changes of volume greater than those which occur in normal cement with up to 3 per cent. of this material, even when present to the extent of 18 or 20 per cent. Mr. Messent's views are quoted, and he asserts that the failures at Aberdeen were neither due to the use of cement of defective quality, nor to the practice of regauging and breaking up partially set concrete which had not been resorted to upon this part of the works. He attributes the injuries to the porous nature of the coating of the blocks, and he brings forward the theory of the deposition of magnesia from sea water, its consequent expansion and hence the disruption of the work.

Mr. Carey conducted a series of experiments to test the action of sea water on cement, and for this purpose he made 120 briquettes, some with neat cement and some with 1 part of cement to 2, 3, and 4 parts of sand. An equal number of tests was made of each series, and the whole of them were placed in a wooden cage and secured at the level of low water of ordinary spring tides at the pier-head at Newhaven Harbour forty-eight hours after gauging. They were thus exposed not only to a constantly varying pressure due to the tidal movement, but also to the action of the waves at low water. They were all carefully weighed and submerged for periods which varied as follows:—Fourteen days, twenty-eight days, three months, and six months. They were then surface dried by exposure in the air and weighed afresh. The results are summarised in the following table and are somewhat contradictory, for, as will be seen, there is a slight gain of about 5 per cent. in fourteen days, while at twenty-eight days and at three months there would seem to be a reduction in weight which is made good, however, at the end of twelve months:—

TABLE SHOWING INCREASED WEIGHT OF CEMENT BRIQUETTES.

PROPORTIONS.	14 Days.	28 Days.	3 Months.	6 Months.
	Per Cent.	Per Cent.	Per Cent.	Per Cent.
Neat, . . .	4·98	2·69	1·25	12·06
1 to 1, . . .	4·53	2·28	0·38	9·57
1 to 2, . . .	5·06	2·25	0·51	9·29
1 to 3, . . .	6·16	2·25	2·51	10·2
1 to 4, . . .	4·97	2·42	3·41	12·77

In spite of these fluctuations in the weight, the strength of the briquettes steadily increased with age, though most of the three months' tests show a small falling off; thus equal parts of cement and sand gave on an average 255 lbs. per square inch at fourteen days, 291 at twenty-eight days, 299 at three months, and 348 lbs. at six months. With two parts of sand the briquettes broke at 123, 166, 185, and 242 lbs. respectively. Another series of tests was instituted by Mr. Carey to show the effect upon the strength of the cement of using sea water alone, and diluted with distilled water, or concentrated by partial evaporation.

The following table gives the result, the mean of six tests in each case, using 20 per cent. of water and neat cement:—

TEST BRIQUETTES, $1'' \times 1''$, MADE UP WITH SALT WATER.

Water.	1 month.	2 months.	3 months.	6 months.	9 months.	12 months.
Fresh water,	525	555	750	900	949	678
½ distilled water, ½ sea water,	537	542	690	650	905	592
¼ distilled water, ¾ sea water,	542	550	620	750	950	575
Sea water, 25 per cent. evaporated,	542½	563	590	580	890	605
Sea water, 50 per cent. evaporated,	541	540	800	850	970	495

Mr. Carey states that the cement was not checked or cracked in any way, but in the case of both series of experiments was sound and well burnt. It was so ground as to leave a residue of about 10 per cent. on the 50×50 mesh sieve. The most noticeable feature in the above table is the remarkable effect of using strong brine, as evinced by the excellence of the tests at 3, 6, and 9 months, and the singular falling off in tensile strength of all the samples at the end of the 12 months, pointing, we should think, to some defect in the constitution of the cement. It would have been interesting to know how the cement behaved after the lapse of two or more years.

As Mr. Carey remarks, "the real point at issue is whether the salts of magnesia, which are admittedly deposited from the sea in porous concrete structures, are or are not inert." Arguing from the fact that magnesian limestone when used in large cities, where the rain brings down sulphate of ammonia, is chemically acted upon, and decomposed into magnesium sulphate, which latter is deposited in the form of crystals, and causes disintegration, he advises that in situations where ammonia may be brought into contact with concrete, it is desirable to use fresh water for gauging, and to discard cements containing a high percentage of magnesia. On the other hand, concrete made of sound and well burnt cement, varying from ¼ to ₁⁄₁₂ part by volume and gauged with sea water, has been used for many existing structures in the sea, which have exhibited no signs of deterioration throughout a long term of years. There is, he thinks, "no conclusive evidence to prove that the precipitates from sea water induce disintegration of even fissured or porous concrete" when sound cement has been employed.

In the Aberdeen concrete it is clear that free caustic lime was present and was washed out of the concrete, while magnesia, in the form of magnesium hydrate, was precipitated, leading to the elimination of lime as

14

calcium chloride and sulphate, but we learn nothing more from these facts than that lime is an unstable and soluble body, and that it should not form any considerable proportion of a durable concrete. The experience of Aberdeen has proved in fact this much and no more :—That "an excess of caustic lime or caustic magnesia causes (1) disintegration by the expansion due to hydration, and that (2), being soluble, these substances, when conditions are favourable, may be washed out, leaving the concrete in a honey-combed state.'

Mr. Smith devotes his attention entirely to the works at Aberdeen, and after a brief account of the concrete construction of the docks, he points out that immediately after completion considerable leakage was noticed coming through the concrete walls, and that soon after the entrance walls began to swell, some of the large blocks showing signs of bulging. As the damage appeared to be due to chemical action in the Portland cement, Prof. Brazier, of Aberdeen University, was consulted on the subject, and in a report dated June, 1887, he confirmed this opinion. He states that :—

"The analyses of the series of decomposed cements show a remarkable difference to the original cement, inasmuch as in all these samples there is found a large quantity of magnesia, and a large proportion of the lime in the form of carbonate. I believe this alteration is brought about entirely by the action of sea water upon the cement. There is no other source for either the magnesia or the carbonic acid."

The following table shows the composition of the original cement, and of samples of the decomposed cements :—

CONSTITUENTS.	Original Cement, Dry.	Decomposed Cement from Wing Wall, south side, inside Cofferdam.	Decomposed Cement from Wing Wall, south side, back of Plaster Skin.	Cement from Entrance Wall, above Culverts, inside Dock.	Cement from Entrance Wall, above Culverts, inside Dock, Drippings.	Cement from South Breakwater, west side.
Alumina & oxide of iron,	13·10	26·76	28·42	1·05	1·53	21·60
Silica, ...	20·92	18·04	19·55	1·33	1·31	17·81
Carbonate of lime, .	8·18	6·61	15·78	45·72	35·42	33·08
Hydrate of lime, .	11·26	30·54	16·94	27·85	17·17	..
Hydrate of magnesia,	..	13·57	15·08	21·03	30·96	9·04
Sulphuric acid, .	0·82	2·98	4·23	1·31	0·90	4·89
Magnesia, . . .	0·33	..	..	..	..	12·68*
Soluble in water, .	..	1·50	..	1·71	3·71	..
Caustic lime, . .	45·39	..	..	..	..	..
	100·00	100·00	100·00	100·00	100·00	100·00

We may observe with respect to these analyses that both in the case of the original cement and the deteriorated concrete there are many matters which we are quite unable to reconcile ; the lime is, however, present in the cement in what we must deem to be a dangerous excess.

In addition to these analyses, Prof. Brazier carried out a series of experiments with specimens of the cement digested in various solutions. The cement for these tests was obtained by breaking down a briquette, sifting out and rejecting the fine powder, and reserving for trial pieces about the size of a large-grained shot. In each case, 200 grains of cement tied up in small muslin bags were suspended in 20 ozs. (1 imperial pint) of the various fluids—1. Distilled water. 2. Solution of magnesium chloride

* Carbonate of magnesia.

containing 31 grains, equivalent to 13·05 grains of magnesia, the average proportion of this salt found in sea water. 3. Solution of magnesium chloride of double the above strength, 62 grains to the pint. 4. Solution of magnesium sulphate, containing the same equivalent of magnesia as No. 2. 5. Solution of magnesium sulphate of double the above strength. 6. Sea water, specific gravity 1·026, containing magnesium salts equivalent to 18·37 grains of magnesia per imperial pint. 7. Sea water, duplicate specimen, and a solution of chloride of sodium—250 grains per imperial pint.

The cement samples remained in these various solutions for about three months, and it was then found that No. 1, the distilled water, was but little altered. In No 2, nearly the whole of the dissolved magnesia had been removed, and the clear liquid contained 14·62 grains of lime per pint. No. 3 had also lost nearly the whole of its magnesia, its place being taken by 26·04 grains of lime. Very much the same action went on in the case of the magnesium sulphate solutions 4 and 5, except that the amount of magnesia removed from solution and replaced by lime was relatively smaller. In the sea water tests, again, the magnesia present had been to a great extent precipitated; No. 6 showing a loss of 14·16 grains, and No. 7, 14·72 grains. But little action took place with the salt solution No. 8. These experiments certainly appear to confirm the opinion expressed by Professor Brazier that the magnesium salts in the sea water have a dangerous action upon the lime compounds of the cement in question, but it must be borne in mind that we have no guarantee that the Portland cement itself was originally of good quality, and contained its lime in stable combination with the silica. The analysis of the original cement (which was unfortunately made from a stale sample) shows only that the percentage of lime was very high (63·1 if the water and carbonic acid are eliminated). The disappearance of the silica, iron, and alumina in the fourth and fifth specimens, given in our table on p. 210, is extremely puzzling.

Experiments made by Mr. Pattinson, of Newcastle-upon-Tyne, agreed in all respects with the above, and he suggested certain investigations, subsequently carried out by Mr. Smith, to test the permeability of cement concrete, as it was thought that the sea passing in and out through the porous concrete at every tide would hasten the destructive action. These experiments led Mr. Smith to formulate the following theories, viz.:—That in order to obtain an impermeable concrete (1) the cement must be finely ground; (2) it should not be weaker than 1 of cement to 6 of sand and stones; and (3) the concrete should have sufficient time to set (at least three months) before the application of the pressure, due to a column of sea water.

Professor Brazier, moreover, believed that the observed expansion of the concrete in question was due to the substitution of magnesium hydrate and calcium sulphate, for the lime therein originally combined with carbonic acid. Mr Smith states apparently on his authority—"There is no doubt that in permeable concrete the whole of the lime forms a base on which carbonic acid and the salts contained in the sea water act chemically, that the resulting deposit occupies a larger space than the lime base, and that the force of chemical action exceeds the adhesive power of the cement, thus effecting the rupture of the mortar, and expansion of the mass of concrete."

It will be evident that a source of danger is herein indicated if we admit that the cement compounds, consisting of hydrated silicates of lime, not to any great extent of free lime, or carbonate of lime, as assumed by Mr. Smith, are liable to be attacked and damaged by the salts found in sea water. Numerous instances are, however, on record where cement concrete has withstood for a long series of years the influence of sea water, not only under a great tidal range, but under the still more trying action of alternate periods of submersion in the sea, and exposure to the air. In many of the great dock and harbour works we have examples of Portland cement concrete, which for upwards of forty years have resisted **most** perfectly the attacks **of sea** water, and there can be but little doubt in the minds of all **unbiassed** observers that in a well-made sample of Portland cement, when once properly indurated as a hydrated double silicate of **lime** and alumina, we possess a material quite unacted **upon by sea** water. The failure **of** certain specimens of concrete **can, we think, in** every case be assigned either to mistakes in the manufacture, **defects in** the quality, or imperfections in the mode **of** using the **cement.**

Appendix D.—Keates' Double Bulb Bottle for ascertaining the Specific Gravity of Cement.

This bottle consists of **two bulbs,** the lower somewhat exceeding the upper **in** capacity. The exact capacity of the lower bulb is of no importance. On the neck between the bulbs is a file-mark b, **on** the neck of the upper bulb is a similar **mark** a.

The capacity of **the** upper bulb between the marks a and b must be accurately determined, and may conveniently be either 500 or 1,000 grains in water measure at 60°.

In ascertaining the specific **gravity of** a solid **in small** fragments, small **shot, for** example, the following **is the** mode of procedure :—Fill the bottle

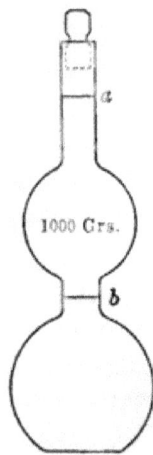

Fig. 30.
Keates' specific gravity bottle.

with distilled water up to the mark b, accurately counterpoise the bottle so filled in a balance, drop the substance (of which the specific gravity is to be taken) carefully and gradually into the bottle until the water rises from b to a, ascertain exactly the weight **of** the material so added ; if **the capacity of** the **upper bulb be 1,000** grains of water, **the weight of the material required to** raise the water **from** b **to** a **is** its specific gravity ; if the capacity of the **bulb be 500 grains of water, the weight** of the substance **added must be multiplied by 2, which will** give the specific gravity.

The **principle of the apparatus is very** simple ; **the** capacity of the upper bulb **is an exact measure** of distilled water, **and when the water is raised** from b **to** a by dropping a **solid into the bottle,** the bulk of that solid equivalent **to the** given **volume of** distilled water is ascertained, **and the** relation between the weights of the **two** is given **by the** weight of the substance added, which **is** either the **specific** gravity direct, if the capacity of the bulb be 1,000 grains, **or** it can be ascertained by multiplying the weight **of** the solid by the number which represents the **part of** 1,000 represented by the capacity of the bulb, &c.

If the solid be soluble in water, any convenient liquid **can be** used in the place **of water** in making the experiment, the only thing necessary being to carefully counterpoise the bottle filled with the liquid up to b in this manner. Petroleum, oil, turpentine, or any liquid suitable to the nature of **the material to be** tested may be used, all other things remaining the same.

The only precautions to be observed are, that the air, which is apt to cling somewhat to the solid matter when dropped into the liquid, is carefully removed ; and that, if a very volatile liquid be used in the place of water, the bottle should be stoppered or corked to prevent evaporation.

Appendix E.—German Standard Tests.

STANDARDS FOR THE UNIFORM DELIVERY AND TESTING OF PORTLAND CEMENT.

Explanation of What is understood by Portland Cement.—Portland cement is a material resulting from the calcination, carried to the point of incipient fusion, of an intimate admixture of lime and argillaceous substances as its essential components ; such calcination being followed by the grinding of the product to the fineness of flour.

I. PACKAGES AND WEIGHT.

Portland cement must, as a rule, be packed in standard-sized casks weighing 180 kilograms gross, and 170 kilograms nett (397 lbs. and 375 lbs.), and in half-sized standard casks of 90 kilograms gross, and 85 kilograms nett (199 and 188 lbs.). The gross weight must be clearly marked on the casks.

If the cement is wanted in other sized casks or in sacks the gross weight must be plainly indicated in figures upon each package.

Leakages, as also variations in individual packages, must not exceed 2 per cent. The casks and sacks must bear, in addition to the weights, the name of the firm or the factory trade mark of the maker plainly indicated on each.

Facts in support of I.—Both in the interests of the purchaser, and also to ensure sound business, the adoption of a uniform weight (for packages) is urgently required. For this purpose, by far the most usual, and as respects the commercial world, the almost universal weight of 180 kilograms tare = about 400 lbs. English has been selected.

II. TIME NEEDED TO SET.

In accordance with the purpose for which it is required Portland cement may be furnished to set either slowly or rapidly.

Such cements are to be regarded as slow-setting, which take two hours and more to become set.

Explanations to II.—In order to determine the period needed to set, in the case of slow-setting cements, a sample of the neat cement is stirred for three minutes with water into the form of a stiff paste, but for quick setting cements the stirring is to be for one minute only. The mixture must then be spread on a glass plate at a single operation in the form of a pat about 1½ centimetres (½ inch) thick, but thinning out at the edges. The consistency of the cement paste needed for this pat is to be such that when placed upon the plate, with a spatula, a few taps upon the glass suffice to cause the mass to flow out to a thin edge ; for this purpose from 27 to 30 per cent. of water will generally suffice. As soon as the pat has become hard enough to withstand a gentle pressure with the finger-nail the cement is to be regarded as having become set. In order to ascertain precisely the setting time, and to determine the beginning of the set, which (as the cement must be worked up before the set begins) is a matter of importance in the case of quick-setting cements, a normal needle of 300 grams (10½ oz.) weight is employed, which has a flat end with a superficial area of 1 square millimetre (0·015 square inch). This surface being at right angles to the axis, a metal ring of 4 centimetres in height and 8 centimetres in clear diameter (1½ inches and 3 inches) is placed on a glass plate, filled with cement paste, gauged as above, and placed beneath the needle. The interval of time which elapses between the gauging and that when the needle is no longer able to penetrate the entire depth of the paste, is regarded as that needed for the beginning of the set, and the time which expires until the needle will no longer produce any noteworthy impression on the hardened surface is the time required for the set. As the setting of cement is influenced by the air-temperature and by that of the water used for gauging purposes, in that the higher is the temperature the more rapid is the set, and the lower the temperature the longer is the time needed for setting, it is advisable, in order to obtain uniformity in the results, to conduct the observations at a mean temperature of air and water of from 15° to 18° Celsius (59° to 64·4° F.).

During the set there should be no perceptible evolution of heat in the case of slow-setting cements, while, on the other hand, quick-setting cements may manifest a notable increase in temperature.

Portland cement becomes by long storage slower in set, and increases in tensile strength, if it be kept in dry places, free from draughts. The still widely prevailing opinion that Portland cement deteriorates in quality by long keeping is, therefore, erroneous, and contract clauses which specify the employment only of fresh cement should be abandoned.

III. Freedom from Change in Volume.

Portland cement must be constant in volume. The decisive test of this quality shall be that a pat of neat cement, placed on a glass plate and protected from too rapid drying, immersed in water at the end of twenty-four hours, and kept under constant observation for a longer period, shall manifest no tendency to crumble or to show signs of splitting at the edges.

Explanations to III.—In carrying out this test, the pat prepared to determine the time needed for setting should, after the lapse of twenty-four hours in the case of slow-setting cements, but in any case only after the pat has become set, be placed in water. In the case of quick-setting cements, the pats may go into water in an even shorter period of time. The pats of slow-setting cement must, until they shall have become set, be protected from draughts and sunshine; this may best be done by putting them into a covered box, or by placing them beneath a damp cloth. By this means the formation of hair-cracks, due to shrinkage in rapid drying, will be avoided; such cracks generally appear in the centre of the pat, and are liable to be mistaken by unskilled observers for cracks caused by blowing.

If during the process of induration under water any crumbling takes place, or if cracks become manifest, this undoubtedly denotes the blowing of the cement; that is to say, that in consequence of its volumetric expansion the cement becomes cracked and a gradual disintegration of the first-formed cohesion takes place, which may ultimately lead to the complete disruption of the mass. The indications of blowing generally become apparent in the pats after about three days, but at all events observations extending over twenty-eight days are sufficient.

IV. Fineness of Grinding.

Portland cement is to be so finely ground that the residue from a sample of the same passed through a sieve of 900 meshes per square centimetre (5,806 per sq. inch or 76 per lin. inch) shall not exceed 10 per cent. The diameter of the sieve wire should be equal to half the clear opening of the mesh.

Explanations to IV.—In the case of each sieve test 100 grams (0·22 lb.) of cement are to be used.

As cement is in nearly every instance employed with sand, and in some cases with large proportions of sand, while the cohesive strength of a sample of mortar is greater just in proportion as the cement is finely ground (because under these circumstances more particles of the cement come into action), the fineness of the grinding is of the utmost importance. It appears, therefore, to be evident that the fineness of the particles should be uniformly checked by the use of a fine sieve of the above number of meshes to the inch. It would, however, be misleading if the quality of a cement were to be determined alone by the fineness of the grinding, for inferior tender-burned cements are much oftener found to be well ground than good hard-burnt cements. The latter will, however, even if coarsely ground, possess a greater cohesive strength than the former. If the cement is to be used mixed with lime, it will be expedient to employ hard-burnt cements, extremely well ground, the relatively high cost of which will be compensated for by the manifest improvement of the mortar.

V. Test to Ascertain the Strength of the Cement.

The cohesive strength of Portland cement must be ascertained by testing a mixture of cement and sand. The testing should take place upon a uniform system both in tension and compression, and should be carried out with test samples of the same shape and sectional area, and with similar apparatus.

Together with these tests it is expedient also to ascertain the cohesive strength of the neat cement.

The tensile tests are to be carried out with briquettes, having a sectional area at the point of fracture of 5 square centimetres (0·775 square inch); in the tests for compression **cubes** should be used, having sides 50 square centimetres **in area** (7·75 square inches).

Facts in **support** *of V.*—As it is impossible uniformly to pronounce an opinion upon the cohesive strength of a sample of cement when employed with sand by means of experiments conducted with that cement without sand, particularly when it becomes necessary to compare the cements of different makers, it is, therefore, imperative to test the cement along with sand.

The testing of neat cement is to be advocated when it is required to prove the value of Portland cement as compared with mixed cements and other hydraulic binding materials, because on account of its high tensile strength when used **neat,** in which respect it surpasses all other similar materials, it shows to greater **advantage** tested thus than when tested with sand.

Although **the** proportional strength under compression and under tension differs in the case of various binding agents, it is, nevertheless, customary to employ the tensile strength as indicative of the value of the different hydraulic cements. This, however, leads to **an** unfair valuation of their properties. And further, whereas in all practical applications of mortar the resistance to compression is the first consideration, it is only these latter tests which can be relied upon to furnish really conclusive results.

To secure the necessary uniformity in the tests it is advisable to make use of the same description of testing apparatus and tools as those employed in the Royal Testing Station at Charlottenburg, Berlin (see Fig. 24, p. 142).

VI. Tests under Tension and Compression.

Slow-setting Portland **cement** shall, when tested with 3 parts **by** weight of standard sand, after a period of twenty-eight days induration—one day in air and twenty-seven days in water—attain a minimum tensile strength of 16 kilograms per square centimetre (227 lbs. per square inch). The strength under compression shall be at least 160 kilograms per square centimetre (2,275 lbs. **per** square inch). In the case of quick-setting cements **the** tensile strength after twenty-eight days is generally inferior to the **above.** It is, therefore, necessary, when stipulating **the** amount of tensile strength, to indicate also the time needed for the set.

Explanations.—As different cements may vary very considerably as **respects their power to unite** sand, upon which property their industrial **application mainly depends, it** is imperatively necessary, in the comparison **of several samples of cements, to** test them with a large proportion of sand. **It is assumed that the most** suitable proportion for this purpose is 3 parts by weight **of sand to** 1 part of cement. Since it is with 3 parts of sand that **the** binding power of **the** cement comes adequately into play.

Cement which exhibits a greater relative tensile strength as compared with the crushing strain often manifests a superior capacity for uniting sand, and merits for this reason, as likewise for its greater strength with an equal amount of **sand, to** command an increased selling price.

The standard **test under** compression is that after twenty-eight days, since **it is not** possible when comparing several kinds **of** cements to properly **ascertain** the binding capacity in **a** shorter period of time. Thus, for **instance,** after twenty-eight days the tests **of** different samples of cements **in** compression **may** be **alike,** whereas, after seven days only, there were important differences. **The tensile** strength **at** twenty-eight days constitutes the decisive test on **delivery** of the cement. **If,** however, **a** check-test at seven days is needed, **this may** be secured by a preliminary test when the ratio of tensile strength **after** seven days to that at twenty-eight days in the case of any given sample of cement has been determined. This preliminary test may also be carried out with neat cement, when the proportionate strength of the neat cement to that with 3 parts of cement at twenty-eight days has been ascertained.

It is advisable in all **cases,** when there **is a** possibility of so doing, to **extend the** tensile tests **over a** longer **period,** by means of cement samples **prepared beforehand for** this purpose, **in order to learn** the behaviour of **various cements during a** longer period **of induration.**

To attain uniformity in the results **it is in all** cases necessary to employ **sand having grains of the same size** and of **the** same quality. This normal **sand is prepared as follows :**—The purest possible quartz sand is washed, **dried, and passed through a** sieve of 387 meshes per square inch to **eliminate the coarser particles,** and then through one of 774 meshes per **square inch to remove the finer** particles. The thickness of the wires not **to exceed 0·015 inch and 0·012** inch respectively.

As all samples of quartz sand when treated **in** the same way do not manifest **the same cohesive strength,** it is necessary to ascertain whether the **specimen of normal sand** employed gives uniform tensile tests with those **shown in the case of the** sand supplied for testing purposes under the authority **of the** German Cement Manufacturers' Association, which sand is also used **at the** Royal Testing Station at Charlottenburg, Berlin.

Description of the Tests employed to ascertain the Tensile and Cohesive Strength.—As **it is** important to secure absolute uniformity in the results in testing **the** same cement in different places, it is imperatively necessary **to** adhere most strictly to the rules that are here indicated. In order to obtain correct averages at least ten test pieces must be prepared for each set of experiments.

PREPARATION OF THE CEMENT WITH SAND TEST-PIECES.

Tensile Tests.—The tensile test-pieces may be prepared either by hand or by mechanical means.

(a) *By Hand Work.*—On the sheet of metal, or thick glass, serving for the formation of the test-pieces, must be placed five pieces of blotting-paper saturated with water, and on these are laid the five moulds, also rinsed out with water. Then 250 grams (8·8 ozs.) of cement and 750 grams (26·4 ozs.) of dry normal sand are weighed out, and well mixed together in a basin. To this mixture must now be added 100 c.c. = 100 grams (3·52 ozs.) of pure fresh water, and the whole mass is then briskly stirred for five minutes. The five moulds are then filled with the mortar thus prepared, which is pressed into them, and well heaped up into a curved form above the top. By means of an iron spatula, having a surface of $2\frac{1}{2} \times 1\frac{3}{4}$ inches, 13 inches in length, and weighing $7\frac{1}{4}$ ozs., the mortar is tapped softly near the edges to begin with, and then gradually harder, so as to drive it down into the mould, until the whole mass becomes elastic, and water appears on the surface. It is absolutely necessary to continue the process of the filling-in of the moulds until this stage is reached, which will need about one minute per mould. It will not do to introduce a second supply of mortar and to beat that down, because it is essential to keep the test-pieces made with the same kind of cement at different testing stations identical in point of density. The superfluous cement is then struck off with a knife, and the surface is polished over with the same. Then the mould is carefully removed, and the test-pieces are placed in a box, lined with zinc and provided with a cover, in order to avoid unequal desiccation, caused by variations in the temperature. Twenty-four hours after they are made, the test-pieces must be placed in water, and care must be taken to keep them covered with water during the whole period of induration.

(b) *Preparation by Mechanical Means.*—After the mould, with the filling box attached to it, has been securely fastened to the bottom plate by means of the two thumb-screws, 180 grams of the mortar, prepared as in the foregoing instructions (a), are introduced into the mould for each test, and the iron moulding core is placed in position. Then, by means of Dr. Böhme's percussion apparatus, having a hammer weighing 4·4 lbs., 150 blows are imparted to the core.

After the removal of the filling-box and the core, the test-piece is struck off level and polished as before; then taken off in the mould from the bottom plate, and dealt with as in the case of (a), above.

By careful attention to the foregoing rules, hand-made test-pieces and those prepared mechanically give fairly uniform results. In disputed cases, however, tests made by mechanical means must be regarded as decisive.

Tests under Compression.—In order to secure uniformity in the results of testing by compression carried out at various testing stations, it is necessary to have recourse to mechanical preparation of the tests.

Fourteen ounces of cement are weighed out and well mixed in a basin with 42 ozs. of dry normal sand. 160 c.c. = 5·63 ozs. water are then added, and the compound is briskly stirred into mortar for five minutes. Then 30 ozs. of this mortar are inserted into the cubical mould, having the filling-box attached, the whole being securely fastened by screws to the bottom plate. The iron core is placed in the mould, and is caused to receive 150 blows from Dr. Böhme's percussion apparatus with a hammer weighing 4·4 lbs. After the removal of the filling-box and the core, the test-block is struck off

and polished, taken off the bottom plate in the mould, and dealt with as in
the case of (a), above.

Preparation of Test-Pieces of Neat Cement. —The moulds are slightly
oiled on the inner surface, and placed on the metal or glass plate (without
any blotting-paper under them). Then 35·2 ozs. of cement are weighed out,
and to this is added 7 ozs. = 200 c.c. of water, and the whole are well
stirred for 5 minutes (this is best done with a pestle), filled into the moulds,
so as to be well heaped up above the surface, and then dealt with as in
(a). The moulds can, however, only be removed when the cement is
sufficiently set.

As, during the ramming in of the neat cement, the aim is to produce test-
pieces of the same consistency, the added water must be proportionately
increased in the case of very finely ground or quick-setting cements.

The amount of water employed must always be indicated when stating
the figures representing the tensile strength.

Method of Carrying Out the Tests. —All test-pieces must be tested
immediately they are taken out of the water. As the time the load is
being applied has an influence upon the result, the weight should, in the
case of tensile tests, be increased at the rate of 3·52 ozs. per second. The
mean of ten tests is to be taken as the ascertained tensile strength.

Appendix F.—The Cost of Cement Manufacture.

It may not be wholly amiss in a work of this character if we devote a few
remarks to the cost of the manufacture of cement, a subject of great importance at the present time, and one which has necessarily attracted a
vast amount of attention. There can be no two opinions that many of the
older works are so heavily weighted with capital, representing obsolete
appliances, imperfect processes of manufacture, and money thrown away
upon past failures, that these "dead expenses" severely handicap them in
competing with modern works, constructed upon the most economical lines,
and provided with all the requisite plant for the inexpensive handling of the
raw materials. The cost of the raw materials may in itself constitute a
more or less serious item in the balance sheet, and there can be no doubt
that cement makers able to control their own supply of clay and chalk or
limestone in the immediate vicinity of their works have a great advantage
over those who have to bring their materials from a distance, or to obtain
them at second hand. In certain parts of this country remote from the
chalk formation, as, for instance, in the Tyne Districts, where fuel is
relatively cheap, cement factories have been established, which depend
for their supply upon chalk brought back as ballast in colliers. But these
works find themselves severely handicapped in the competition with factories
established on the Thames and Medway, close to the quarries. It may be
taken for granted that most of the older works, which, as we have seen,
cluster round the chalk districts south-east of London, where both Medway
clay and chalk are abundant, can command unlimited supplies of raw
materials at a price ranging from 2s. to 2s. 6d. per ton of finished cement.
Even in those works where the chalk has to be bought, it can usually be
laid down on the works at from 1s. 2d. to 1s. 6d. per ton ; the clay costing
in a similar way from 2s. 2d. to 2s. 8d. per ton. In the lias cement
districts in the Midland Counties the cost of raw materials is relatively
high, owing to the increased cost of quarrying the thin beds of stone under
great depths of shale and clay. We know of one cement works successfully
conducted in the Midlands, where the cost of the raw materials from the

lias formation delivered on the site is set down at **4s. 6d.** per ton. **Even** in the districts where the hard limestone takes the place of chalk, the stone can be quarried and laid down on **the works for** from **1s. 3d. to 1s. 9d.** per ton. The price of fuel (coke or coal) is also an **important consideration to** the cement maker, as this item **amounts to** about one-quarter of the total cost of manufacture, and cheap **coke may** outweigh the disadvantages due to relatively costly works **or expensive raw** materials. The carriage **of** Portland cement on the **rail is another serious item, as** most of the companies are in the habit **of charging very high** rates for cements, as compared with coal or stone. It **may be taken as a** general rule that no large works can be advantageously **carried on, which** have **not direct** communication with the rail by **means of sidings, or which do not** possess cheap water communication. The **export trade in Portland is** very extensive, but this business **is almost entirely in the hands of makers on the** Thames and Medway, or in the **sea-port towns. The following estimates** will give some idea of the items **of expenditure under various heads in** the manufacture of Portland cement.

The following analysis of the cost of cement making on the Medway was published in the *Portland Cement Journal* in 1892. It purports to give the exact figures "taken from actual accounts for the years 1890 and 1891." There is no provision made for interest on capital, which is stated to have been over £40,000 for buildings and plant, and represents rather a large, but by no means an unusual expenditure for this output. The property is freehold, and the sum of £500 charged for depreciation only relates to the depreciation on fixed machinery and plant. No sum appears to be set down for agency or travelling, bad debts, and incidental losses, and it is stated with respect to cost of chalk, clay, coal, and coke, that the figures given do not include labour thereon after reaching the wharf, such labour being included in the wages item:—

	1890.	Cost per ton of finished cement.	1891.	Cost per ton of finished cement.
Quantity of cement made in tons, . .	21,336		20,175	
	£	Shillings.	£	Shillings.
Chalk and freight, ...	2,326	2·180	2,571	2·549
Clay and freight, ...	593	·556	601	·686
Coal and freight, ...	4,935	4·626	5,980	5·928
Coke and freight, ...	5,037	4·722	5,562	5·513
Wages, ...	6,457	6·053	6,408	6·353
Rates and taxes, ...	170	·159	170	·168
Repairs, materials, and wages (on repairs),	1,197	1·123	1,761	1·726
Stores, oil, grease, firebricks, iron, steel, &c., not charged to repairs, . .	1,133	1·063	1,151	1·141
Trade charges, postage, stationery, &c., .	86	·081	93	·092
Insurance, ...	85	·081	79	·078
Depreciation, amount written off fixed machinery and plant, . . .	500	·408	500	·495
Manager, ...	250	·234	250	·248
Books kept in London, salaries, stationery, &c., proportion charged, . . .	195	·183	218	·216
Add to wages, say, for cement in casks, 5,000 tons at 1s. 4¾d., . . .	£22,964	21·529 = 1·396	£25,436	25·193 1·396
		22·925 = 23/.		26·589 = 26/7

From our experience of the cost of raw materials in the London district, we regard the average cost of chalk and clay on the wharf for the two years, amounting to 2·985s., or, as nearly as possible, 3s. per ton of finished cement, as rather too high, and the charge for fuel is certainly extravagant, amounting as it does to over 10s. 4d. per ton of cement. It will be noticed that the prices of 1891 are more than 25 per cent. higher than those of 1890. **It** is stated that the average selling price of the cement at the works was 28s. 6d. per ton in 1892, and it will thus be seen that the manufacture was being conducted at an actual loss, if proper provision is made for the items **to** which we have called attention. In the face of the above figures, it can easily be understood that the Portland cement trade is **in** a very bad **state,** and that **many** of the large concerns are paying no dividend.

It will readily be understood that the publication of an estimate of this kind **evoked** considerable criticism, and shortly afterwards another estimate was published which emanated from a correspondent, who stated that he could erect works in the Medway district, on certain new lines, capable of producing 35,000 tons of cement a year, **at** a cost of 16s. 9d. per ton. He proposed **to** erect the manufactory in question on freehold land, to be acquired on advantageous terms, the **expense of** which land, together with the outlay **on the** erection **of** the **necessary** buildings was not to exceed £20,000. **This sum** does not appear **to** include the cost of the machinery and plant. **It will be seen** that in **this** analysis, while the cost of the raw materials **delivered on** works is 2s. **8d.** per ton of cement, that of the fuel amounts **only to** 5s. 6·3d., as compared with 3s. and 10s. 4d. respectively, the mean **of the** figures previously given in each **case.** The wages are about the same **in each** estimate.

Cost of manufacture of 35,000 tons of cement on the Medway.		Materials.		Wages.
52,500 tons of chalk alongside wash-mill, .	1/4	£3,062 10 0	..	..
17,500 tons of clay alongside wharf and landing, ...	1/10	1,604 3 4	/4	£291 13 4
70,000 tons into wash-mill, ...	..	..	/3	875 0 0
1,650 tons of Welsh coal delivered, .	20/	1,650 0 0	/4	27 10 0
13,000 tons of coke and faggots, . .	12/4	8,016 13 4	/8	433 13 4
Repairs and renewals, ...	/8	1,166 13 4	/4	583 6 8
Stores, oil, grease, timber, and other sundry goods, ...	/6	875 0 0	..	..
Burning—including filling 1/6, drawing /6, .	..	..	2/	3,500 0 0
Miller wet /2, dry /4, ...	..	..	/6	875 0 0
Sundries—(labour) engineer 66/, man 30/, drivers 60/, stoker 25/, sampler 30/, watchman and house 20/, tallow 25/, two odd men 40/, bags 20/, one night driver 30/, stoker 25/, man 40/. Total £20, 5/ per week for 52 weeks, ...	..	..	..	1,053 0 0
Casks and sacks, filling, heading, and loading, ...	..	..	1/4	2,333 13 4
Rates and taxes /2, trade charges and stationery /1½, ...	/3½	510 8 4	..	..
Insurance—Boiler, fire, and liability, .	/1	145 16 8	..	..
Depreciation fund, fixed buildings and plant,	/4	583 6 8	..	..
Manager and clerk, ...	..	..	..	500 0 0
Allow per contingencies, ...	/3	437 10 0	/6	875 0 0
Materials = 10/3·78 per ton. Wages = 6/5·81 „		£18,052 1 8		£11,347 16 8
				18,052 1 8
Total, . 16/9·79 „				£29,399 18 4

By way of comparison with these estimates, we insert an analysis of the cost of producing cement by the semi-dry process in a works in the lias district. The stone here is relatively costly, but fuel is cheaper than nearer London. In consequence of the heavy outlay incurred upon the repair of the works, which were kept in a high state of efficiency, by constant renewals, this item appears excessive. It will be seen that the total cost of manufacture after the repayment of 5 per cent. on a rather heavy capital was under 25s. per ton, or deducting interest 22s. 9d. per ton, and this is about the figure we should set down for a well managed works in the south of England. We understand, on the Continent, where fuel is dear and where the cement is very well ground, that in works in which a hard limestone is used the expense of making Portland cement can be kept well within these figures.

COST OF PORTLAND CEMENT MANUFACTURE IN THE MIDLANDS.

Capital for land, buildings, and plant, £21,000.
Amount of cement made in tons, 9,960.

	£	s.	d.	s.	d.
Raw material—limestone, shale, and royalty, . .	1,867	10	0	3	9
Coal and coke, ...	2,622	0	0	5	3·2
Oil, grease, and small stores, ...	540	0	0	1	1
Wages, ...	3,590	0	0	7	2·5
Management, ...	525	0	0	1	0·6
Agency, advertisements, &c., ...	282	0	0	0	6·8
Repairs, materials, and wages on repairs, . .	1,320	0	0	2	8
Office printing, stationery, ...	190	0	0	0	4·5
Rates and taxes, ...	70	0	0	0	1·7
Bad debts, ...	27	10	0	0	0·6
Interest on capital at 5 per cent., ...	1,050	0	0	2	1·3
Depreciation on machinery and plant, . . .	225	0	0	0	6·3
Insurance—buildings, boilers, and workmen, . .	56	0	0	0	1·3
Cost = nearly 25s. per ton of cement.	£12,365	0	0	24	10·8

Per ton of cement.

Appendix G.—The Portland Cement Trade of the World.

Mr. Giron recently read before the Engineers' Club at Philadelphia a paper on the trade of the world in Portland cement, in the course of which he said that the present annual production in Europe amounts to over 20,000,000 barrels, and its commercial value to over $36,000,000. The first factory was established at Northfleet, on the Thames. The process was so crude that in 1850 only four factories were in operation. In England there are now over 8,300,000 barrels made each year. The process is much the same as it was twenty years ago. The raw materials are chalk and clay, both very pure, and although inferior processes are employed they make a satisfactory cement. A few years ago the entire product of the kilns was put on the market, but the fineness of the Continental cements led English makers to improve their processes, although even now English cement is not, as a rule, as fine as German or French Portland. The manufacture was first introduced into Germany in 1852. Now there are sixty large works, having the same annual production as

England. The raw materials are of exceedingly unfavourable character, but the manufacturers have made a serious study of the properties and uses of Portland cement, and now know exactly what rules to follow to regulate their operations, and the consumer can depend on the product offered him. The association of manufacturers has had much to do with the immense development of the industry in Germany. In France the industry grew slowly, the total production in 1880 hardly exceeding 750,000 barrels a year; now it is 1,800,000 barrels. The works of the French Cement Company at Boulogne form the largest Portland cement factory in the world, turning out about 800,000 barrels a year. In Russia the first works were established in 1857, and there are now eight factories producing 900,000 barrels a year. In Belgium there are four works producing 800,000 barrels. In Italy the Portland cement industry does not really exist, although a certain kind of natural Portland is made. In Denmark and Norway and Sweden there are ten factories making about 800,000 barrels. Portland cement was imported into the United States as early as 1868. In 1882 the amount imported was 370,406 barrels, and last year it exceeded 3,000,000 barrels. Little effort has been made to develop the manufacture there. The materials for manufacture are as difficult to handle as in Germany, and the processes are similar. The process requires about eight days and demands great care to produce a uniform product. The cost of the system is too great to make it successful in America. The process used in the Portland cement works at Coplay in Pennsylvania has entirely revolutionised the science of cement making. At that factory the raw compound is burnt in a powdered condition while travelling in an inclined rotary furnace in an intensely hot petroleum flame, and a few hours is sufficient to finish the process. The cement is guaranteed to stand 400 lbs. in seven days, 500 lbs. in a month, and 600 lbs. in three months, and to leave no more than 10 per cent. residue on a No. 80 sieve.

Appendix H.—Dorking Stone Lime.

Writers on building materials and the authorities who are responsible for all our most carefully drawn specifications have, with a surprising degree of unanimity, concurred in calling the lime prepared from the lower chalk in the vicinity of Dorking a "stone lime." We have sought for some explanation of this fallacy, and offer the following suggestion:— Certain of the beds near Betchworth, and in the chalk hills eastward of Dorking, are much indurated and discoloured with iron salts, and a bed just between the chalk and the gault, known as firestone, is very hard, and is locally used for building. This indurated chalk is rejected by the lime burners as, owing to the high percentage of silica and alumina present therein, it yields a poor and very sluggish lime, but the raw material has certainly some pretensions to be called a stone. The lime prepared round Dorking is all burnt from chalk which contains from 5 to 10 per cent. of clay, and is of a pleasant buff tint, and as we have seen there is a bed of stone below it. Even Bartholomew, writing in 1840, is careful to point out that the "stone-lime" of Dorking should be specified, and his precepts are still diligently followed at the present day by the conscientious architect who knows what he wants for the preparation of a good sample of mortar.

INDEX.

.15

BELL AND DAIN, LIMITED, PRINTERS, MITCHELL STREET, GLASGOW.

SECOND EDITION, *Revised and Enlarged.*
In Large 8vo, Handsome cloth, 34s.

HYDRAULIC POWER

AND

HYDRAULIC MACHINERY.

BY

HENRY ROBINSON, M. Inst. C.E., F.G.S.,

FELLOW OF KING'S COLLEGE, LONDON; PROF. OF CIVIL ENGINEERING,
KING'S COLLEGE, ETC., ETC.

𝕎ith numerous 𝕎oodcuts, and 𝕊ixty=nine ℙlates.

GENERAL CONTENTS.

Discharge through Orifices—Gauging Water by Weirs—Flow of Water through Pipes—The Accumulator—The Flow of Solids—Hydraulic Presses and Lifts—Cyclone Hydraulic Baling Press—Anderton Hydraulic Lift—Hydraulic Hoists (Lifts)—The Otis Elevator—Mersey Railway Lifts—City and South London Railway Lifts—North Hudson County Railway Elevator—Lifts for Subways—Hydraulic Ram—Pearsall's Hydraulic Engine—Pumping-Engines—Three-Cylinder Engines—Brotherhood Engine—Rigg's Hydraulic Engine—Hydraulic Capstans—Hydraulic Traversers—Movable Jigger Hoist—Hydraulic Waggon Drop—Hydraulic Jack—Duckham's Weighing Machine—Shop Tools—Tweddell's Hydraulic Rivetter—Hydraulic Joggling Press—Tweddell's Punching and Shearing Machine—Flanging Machine—Hydraulic Centre Crane—Wrightson's Balance Crane—Hydraulic Power at the Forth Bridge—Cranes—Hydraulic Coal-Discharging Machines—Hydraulic Drill—Hydraulic Manhole Cutter—Hydraulic Drill at St. Gothard Tunnel—Motors with Variable Power—Hydraulic Machinery on Board Ship—Hydraulic Points and Crossings—Hydraulic Pile Driver—Hydraulic Pile Screwing Apparatus—Hydraulic Excavator—Ball's Pump Dredger—Hydraulic Power applied to Bridges—Dock-gate Machinery—Hydraulic Brake—Hydraulic Power applied to Gunnery—Centrifugal Pumps—Water Wheels—Turbines—Jet Propulsion—The Gerard-Barré Hydraulic Railway—Greathead's Injector Hydrant—Snell's Hydraulic Transport System—Greathead's Shield—Grain Elevator at Frank-fort—Packing—Power Co-operation—Hull Hydraulic Power Company—London Hydraulic Power Company—Birmingham Hydraulic Power System—Niagara Falls—Cost of Hydraulic Power—Meters—Schönheyder's Pressure Regulator—Deacon's Waste-Water Meter.

"A Book of great Professional Usefulness."—*Iron.*

. The SECOND EDITION of the above important work has been thoroughly revised and brought up to date. Many new full-page Plates have been added—the number being increased from 43 in the First Edition to 69 in the present. Full Prospectus, giving a description of the Plates, may be had on application to the Publishers.

LONDON: CHARLES GRIFFIN & CO., LTD.; EXETER STREET, STRAND.

A CATALOGUE

OF

SCIENTIFIC AND TECHNICAL WORKS

PUBLISHED BY

CHARLES GRIFFIN & COMPANY,

LIMITED.

MESSRS. CHARLES GRIFFIN & COMPANY'S PUBLICATIONS may be obtained through any Bookseller in the United Kingdom, or will be sent direct on receipt of remittance to cover published price. To prevent delay, Orders should be accompanied by a Remittance. Postal Orders and Cheques to be crossed "SMITH, PAYNE & SMITHS."

☞ GENERAL and MEDICAL CATALOGUES forwarded Post-free on Application.

LONDON:
EXETER STREET, STRAND.

INDEX.

THE DESIGN OF STRUCTURES:

A Practical Treatise on the Building of Bridges, Roofs, &c.

BY S. ANGLIN, C.E.,

Master of Engineering, Royal University of Ireland, late Whitworth Scholar, &c.

With very numerous Diagrams, Examples, and Tables.
Large Crown 8vo.   Cloth, 16s.

The leading features in Mr. Anglin's carefully-planned "Design of Structures" may be briefly summarised as follows :—

1. It supplies the want, long felt among Students of Engineering and Architecture, of a concise Text-book on Structures, requiring on the part of the reader a knowledge of ELEMENTARY MATHEMATICS only.

2. The subject of GRAPHIC STATICS has only of recent years been generally applied in this country to determine the Stresses on Framed Structures ; and in too many cases this is done without a knowledge of the principles upon which the science is founded. In Mr. Anglin's work the system is explained from FIRST PRINCIPLES, and the Student will find in it a valuable aid in determining the stresses on all irregularly-framed structures.

3. A large number of PRACTICAL EXAMPLES, such as occur in the every-day experience of the Engineer, are given and carefully worked out, some being solved both analytically and graphically, as a guide to the Student.

4. The chapters devoted to the practical side of the subject, the Strength of Joints, Punching, Drilling, Rivetting, and other processes connected with the manufacture of Bridges, Roofs, and Structural work generally, are the result of MANY YEARS' EXPERIENCE in the bridge-yard ; and the information given on this branch of the subject will be found of great value to the practical bridge-builder.

"Students of Engineering will find this Text-Book INVALUABLE."—*Architect.*

"The author has certainly succeeded in producing a THOROUGHLY PRACTICAL Text-Book."—*Builder.*

"We can unhesitatingly recommend this work not only to the Student, as the BEST TEXT-BOOK on the subject, but also to the professional engineer as an EXCEEDINGLY VALUABLE book of reference."—*Mechanical World.*

"This work can be CONFIDENTLY recommended to engineers. The author has wisely chosen to use as little of the higher mathematics as possible, and has thus made his book of REAL USE TO THE PRACTICAL ENGINEER. . . . After careful perusal, we have nothing but praise for the work."—*Nature.*

LONDON: EXETER STREET, STRAND.

ASSAYING (A Text-Book of):

For the use of Students, Mine Managers, Assayers, &c.

By C. BERINGER, F.C.S.,
Late Chief Assayer to the Rio Tinto Copper Company, **London,**

And J. J. BERINGER, F.I.C., F.C.S.,
Public Analyst for, and Lecturer to the Mining Association of, Cornwall.

With numerous Tables and Illustrations.   Crown 8vo.   **Cloth,** 10/6.
Third Edition; Revised.

GENERAL CONTENTS. — PART I. — INTRODUCTORY: MANIPULATION: Sampling: Drying; Calculation of Results—Laboratory-books and Reports. METHODS. Dry Gravimetric; Wet Gravimetric—Volumetric Assays: Titrometric, Colorimetric, Gasometric—Weighing and Measuring—Reagents—Formulæ, Equations, &c.—Specific Gravity.
PART II.—METALS: Detection and Assay of Silver, Gold, Platinum, Mercury, Copper, Lead, Thallium, Bismuth, Antimony, Iron, Nickel, Cobalt, Zinc, Cadmium, Tin, Tungsten, Titanium, Manganese, Chromium, &c.—Earths, Alkalies.
PART III.—NON-METALS: Oxygen and Oxides: The Halogens—Sulphur and · Sulphates—Arsenic, Phosphorus, Nitrogen—Silicon, Carbon, Boron—Useful Tables.

"A REALLY MERITORIOUS WORK, that may be safely depended upon either for systematic instruction or for reference."—*Nature.*
"This work is one of the BEST of its kind. . . . Essentially of a practical character.
. . . Contains all the information that the Assayer will find necessary in the examination of minerals."—*Engineer.*

PHOTOGRAPHY:

ITS HISTORY, PROCESSES, **APPARATUS, AND** MATERIALS.

Comprising Working Details of all the More Important Methods.

By A. BROTHERS, F.R.A.S.

WITH TWENTY-FOUR FULL PAGE PLATES BY MANY OF THE PROCESSES DESCRIBED, AND ILLUSTRATIONS IN THE TEXT.

In 8vo, Handsome Cloth.   Price 18s.

GENERAL CONTENTS. — PART. I. INTRODUCTORY — Historical Sketch; **Chemistry** and **Optics** of Photography; Artificial Light.— PART II. Photographic Processes.—PART III. Apparatus.—PART IV. Materials.—PART V.—Applications of Photography; **Practical Hints.**

"**Mr.** Brothers has had an experience in Photography so large and varied that any work by him cannot fail to be interesting and valuable. . . . A MOST COMPREHENSIVE volume, entering with full details into the various processes, and VERY FULLY illustrated. The PRACTICAL HINTS are of GREAT VALUE. . . . Admirably got up."—*Brit. Jour. of Photography.*
"For the Illustrations alone, the book is most interesting; but, apart from these, the volume is valuable, brightly and pleasantly written, and MOST ADMIRABLY ARRANGED."—*Photographic News.*
"Certainly the FINEST ILLUSTRATED HANDBOOK to Photography which has ever been published. Should be on the reference shelves of every Photographic Society."—*Amateur Photographer.*
"A handbook so far in advance of most others, that the Photographer must not fail to obtain a copy as a reference work."—*Photographic Work.*
"The COMPLETEST HANDBOOK of the art which has yet been published."—*Scotsman.*

MINE-SURVEYING (A Text-Book of):

For the use of Managers of Mines and Collieries, Students at the Royal School of Mines, &c.

By BENNETT H. BROUGH, F.G.S.,
Late Instructor of Mine-Surveying, Royal School of Mines.

With Diagrams. FOURTH EDITION. Crown 8vo. Cloth, 7s. 6d.

GENERAL CONTENTS.

General Explanations—Measurement of Distances—Miner's Dial—Variation of the Magnetic-Needle—Surveying with the Magnetic-Needle in presence of Iron—Surveying with the Fixed Needle—German Dial—Theodolite—Traversing Underground—Surface-Surveys with Theodolite—Plotting the Survey—Calculation of Areas—Levelling—Connection of Underground- and Surface-Surveys—Measuring Distances by Telescope—Setting-out—Mine-Surveying Problems—Mine Plans—Applications of Magnetic-Needle in Mining—*Appendices.*

"It is the kind of book which has long been wanted, and no English-speaking **Mine Agent** or Mining Student will consider his technical library complete without it."—*Nature.*

"SUPPLIES A LONG-FELT WANT."—*Iron.*

"A valuable accessory to **Surveyors in every department of** commercial enterprise."—*Colliery Guardian.*

WORKS

By WALTER R. BROWNE, M.A., M. INST. C.E.,
Late Fellow of Trinity College, Cambridge.

THE STUDENT'S MECHANICS:
An Introduction to the Study of Force and Motion.

With Diagrams. Crown 8vo. Cloth, 4s. 6d.

"Clear in style and practical in method, 'THE STUDENT'S MECHANICS' is cordially to be recommended from all points of view."—*Athenæum.*

FOUNDATIONS OF MECHANICS.
Papers reprinted from the *Engineer.* In Crown 8vo, 1s.

FUEL AND WATER:
A Manual for Users of Steam and Water.
BY PROF. SCHWACKHÖFER AND W. R. BROWNE, M.A. (See p. 28.)

LONDON: EXETER STREET, STRAND.

PRACTICAL GEOLOGY

(AIDS IN):

WITH A SECTION ON *PALÆONTOLOGY*.

BY

GRENVILLE A. J. COLE, F.G.S.,

Professor of Geology in the Royal College of Science for Ireland.

SECOND EDITION, Revised.   With Illustrations.   Cloth, 10s. 6d.

GENERAL CONTENTS.—PART I.—SAMPLING OF THE EARTH'S CRUST.   PART II.—EXAMINATION OF MINERALS.   PART III.—EXAMINATION OF ROCKS.   PART IV.—EXAMINATION OF FOSSILS.

"Prof. Cole treats of the examination of minerals and rocks in a way that has never been attempted before . . . DESERVING OF THE HIGHEST PRAISE.   Here indeed are 'Aids' INNUMERABLE and INVALUABLE.   All the directions are given with the utmost clearness and precision."—*Athenæum.*

"To the younger workers in Geology, Prof. Cole's book will be as INDISPENSABLE as a dictionary to the learners of a language."—*Saturday Review.*

"That the work deserves its title, that it is full of 'AIDS,' and in the highest degree 'PRACTICAL,' will be the verdict of all who use it."—*Nature.*

"A MOST VALUABLE and welcome book . .   the subject is treated on lines wholly different from those in any other Manual, and is therefore very ORIGINAL."—*Science Gossip.*

"A more useful work for the practical geologist has not appeared in handy form."—*Scottish Geographical Magazine.* .

"This EXCELLENT MANUAL . . . will be A VERY GREAT HELP. . . .' The section on the Examination of Fossils is probably the BEST of its kind yet published. . .   FULL of well-digested information from the newest sources and from personal research."—*Annals of Nat. History.*

SANITARY RULES AND TABLES:

A Pocket-book of Data and General Information useful to Municipal Engineers, Surveyors, Sanitary Authorities, Medical Officers of Health, and Sanitary Inspectors.

BY W. SANTO CRIMP, M.INST.C.E.,

AND CH. H. COOPER, A.M.I.C.E.

LONDON : EXETER STREET, STRAND.

SEWAGE DISPOSAL WORKS:

A Guide to the Construction of Works for **the Prevention of the Pollution by Sewage of Rivers and Estuaries.**

BY

W. SANTO CRIMP, M.INST.C.E., F.G.S.,

Late Assistant-Engineer, London County Council.

With Tables, Illustrations in the Text, and 37 Lithographic Plates. Medium 8vo. Handsome Cloth.

SECOND EDITION, REVISED AND ENLARGED. 30s.

PART I.—INTRODUCTORY.

Introduction.
Details of River Pollutions and Recommenda-
tions of Various Commissions.
Hourly and Daily Flow of Sewage.
The Pail System as Affecting Sewage.
The Separation of Rain-water from the Sewage
Proper.

Settling Tanks.
Chemical Processes.
The Disposal of Sewage-sludge.
The Preparation of Land for Sewage Dis-
posal.
Table of Sewage **Farm Management.**

PART II.—SEWAGE DISPOSAL WORKS IN OPERATION—THEIR CONSTRUCTION, MAINTENANCE, AND COST.

Illustrated by Plates showing the General **Plan and** Arrangement adopted in each District.

Map of the LONDON **Sewage System.**
Crossness Outfall.
Barking Outfall.
Doncaster Irrigation Farm.
Beddington Irrigation Farm, **Borough of**
Croydon.
Bedford Sewage Farm Irrigation.
Dewsbury and Hitchin Intermittent Fil-
tration.
Merton, Croydon Rural Sanitary Authority.
Swanwick, Derbyshire.
The Ealing Sewage Works.
Chiswick.
Kingston-on-Thames, A. B. **C. Process.**
Salford Sewage Works.

Bradford, **Precipitation.**
New Malden, **Chemical Treatment and**
Small Filters.
Friern Barnet.
Acton, Ferozone and Polarite Process.
Ilford, Chadwell, and Dagenham Works.
Coventry.
Wimbledon.
Birmingham.
Margate.
Portsmouth.
BERLIN Sewage Farms.
Sewage Precipitation Works, **Dortmund**
(Germany).
Treatment of Sewage by Electrolysis.

₊ **From** the fact of the Author's having, for some years, had charge of the Main Drainage Works of the Northern Section of the Metropolis, the chapter on LONDON will be found to contain many important details which would not otherwise have been available.

"All persons interested in Sanitary Science owe a debt of gratitude to Mr. Crimp. . . . His work will be especially useful to SANITARY AUTHORITIES and their advisers . . . EMINENTLY PRACTICAL AND USEFUL . . . gives plans and descriptions of MANY OF THE MOST IMPORTANT SEWAGE WORKS of England . . . with very valuable information as to the COST of construction and working of each. . . . The carefully-prepared drawings permit of an easy comparison between the different systems."—*Lancet.*

"Probably the MOST COMPLETE AND BEST TREATISE on the subject which has appeared in our language. . . . Will prove of the greatest use to all who have the problem of Sewage Disposal to face."—*Edinburgh Medical Journal.*

CROMPTON (R. E., V.P.Inst.E.E., M.Inst.C.E.):

DYNAMOS (A Practical Treatise on). With numerous Illustrations. In Large 8vo.

WORKS
By J. R. AINSWORTH DAVIS, B.A.,
PROFESSOR OF BIOLOGY, UNIVERSITY COLLEGE, ABERYSTWYTH.

DAVIS (Prof. Ainsworth): BIOLOGY (An Elementary Text-Book of). In large Crown 8vo, Cloth. SECOND EDITION.

PART I. VEGETABLE MORPHOLOGY AND PHYSIOLOGY. With Complete Index-Glossary and 128 Illustrations. Price 8s. 6d.

PART II. ANIMAL MORPHOLOGY AND PHYSIOLOGY. With Complete Index-Glossary and 108 Illustrations. Price 10s. 6d.

EACH PART SOLD SEPARATELY.

** NOTE—The SECOND EDITION has been thoroughly Revised and Enlarged, and includes all the leading selected TYPES in the various Organic Groups.

"Certainly THE BEST 'BIOLOGY' with which we are acquainted. It owes its pre-eminence to the fact that it is an EXCELLENT attempt to present Biology to the Student as a CORRELATED AND COMPLETE SCIENCE. The glossarial Index is a MOST USEFUL addition."—*British Medical Journal.*

"Furnishes a CLEAR and COMPREHENSIVE exposition of the subject in a SYSTEMATIC form."—*Saturday Review.*

"Literally PACKED with information."—*Glasgow Medical Journal.*

DAVIS (Prof. Ainsworth): THE FLOWERING PLANT, as Illustrating the First Principles of Botany. Large Crown 8vo, with numerous Illustrations. 3s. 6d. SECOND EDITION.

"It would be hard to find a Text-book which would better guide the student to an accurate knowledge of modern discoveries in Botany. . . . The SCIENTIFIC ACCURACY of statement, and the concise exposition of FIRST PRINCIPLES make it valuable for educational purposes. In the chapter on the Physiology of Flowers, an *admirable résumé* is given, drawn from Darwin, Hermann Müller, Kerner, and Lubbock, of what is known of the Fertilization of Flowers."—*Journal of the Linnean Society.*

DAVIS and SELENKA: A ZOOLOGICAL POCKET-BOOK; Or, Synopsis of Animal Classification. Comprising Definitions of the Phyla, Classes, and Orders, with Explanatory Remarks and Tables. By Dr. Emil Selenka, Professor in the University of Erlangen. Authorised English translation from the Third German Edition. In Small Post 8vo, Interleaved for the use of Students. Limp Covers, 4s.

"Dr. Selenka's Manual will be found useful by all Students of Zoology. It is a COMPREHENSIVE and SUCCESSFUL attempt to present us with a scheme of the natural arrangement of the animal world."—*Edin. Med. Journal.*

"Will prove very serviceable to those who are attending Biology Lectures. . . . The translation is accurate and clear."—*Lancet.*

GAS, OIL, AND AIR ENGINES:

A Practical Text-Book on Internal **Combustion Motors** without Boiler.

By BRYAN DONKIN, M. INST. C. E.

With numerous Illustrations. Large 8vo, 21s.

GENERAL CONTENTS.—Gas Engines:—General Description—History and Development—British, French, and German Gas Engines—Gas Production for Motive Power—Theory of the Gas Engine—Chemical Composition of Gas in Gas Engines—Utilisation of Heat—Explosion and Combustion. Oil Motors:—History and Development—Various Types—Priestman's and other Oil Engines. Hot-Air Engines:—History and Development—Various Types: Stirling's, Ericsson's, &c., &c.

"The BEST BOOK NOW PUBLISHED on Gas, Oil, and Air Engines. . . . Will be of VERY GREAT INTEREST to the numerous practical engineers who have to make themselves familiar with the motor of the day , Mr Donkin has the advantage of LONG PRACTICAL EXPERIENCE, combined with HIGH SCIENTIFIC AND EXPERIMENTAL KNOWLEDGE, and an accurate perception of the requirements of Engineers."—*The Engineer.*

"The intelligence that Mr. BRYAN DONKIN has published a Text-book should be GOOD NEWS to all who desire reliable, up-to-date information. . . . His book is MOST TIMELY, and we welcomed it at first sight as being just the kind of book for which everybody interested in the subject has been looking. . . . We HEARTILY RECOMMEND Mr. Donkin's work. . . . A monument of careful labour. . . . Luminous and comprehensive. . . . Nothing of any importance seems to have been omitted."—*Journal of Gas Lighting.*

INORGANIC CHEMISTRY (A Short Manual of).

By A. DUPRÉ, Ph.D., F.R.S., AND WILSON HAKE,

Ph.D., F.I.C., F.C.S., of the Westminster Hospital Medical School.

SECOND EDITION, Revised. Crown 8vo. Cloth, 7s. 6d.

"A well-written, clear and accurate Elementary Manual of Inorganic Chemistry. . . . We agree heartily in the system adopted by Drs. Dupré and Hake. WILL MAKE EXPERIMENTAL WORK TREBLY INTERESTING BECAUSE INTELLIGIBLE."—*Saturday Review.*

"There is no question that, given the PERFECT GROUNDING of the Student in his Science, the remainder comes afterwards to him in a manner much more simple and easily acquired. The work IS AN EXAMPLE OF THE ADVANTAGES OF THE SYSTEMATIC TREATMENT of a Science over the fragmentary style so generally followed. BY A LONG WAY THE BEST of the small Manuals for Students."—*Analyst.*

HINTS ON THE PRESERVATION OF FISH,

IN REFERENCE TO FOOD SUPPLY.

BY J. COSSAR EWART, M.D., F.R.S.E.,

Regius Professor of Natural History, University of Edinburgh.

In Crown 8vo. Wrapper, 6d.

LONDON: EXETER STREET, STRAND.

SECOND EDITION, *Revised.   Royal 8vo.   With numerous Illustrations and*
13 *Lithographic Plates.   Handsome Cloth.   Price* 30s.

BRIDGE-CONSTRUCTION

(A PRACTICAL TREATISE ON):

Being a Text-Book on the Construction of Bridges in Iron and Steel.

FOR THE USE OF STUDENTS, DRAUGHTSMEN, AND ENGINEERS.

BY

T. CLAXTON FIDLER, M. INST. C.E.,

Prof. of Engineering, University College, Dundee.

"**Mr.** FIDLER'S SUCCESS arises from the combination of **EXPERIENCE and** SIMPLICITY OF TREATMENT displayed on every page. . . . **Theory is kept in** subordination to practice, and his book is, therefore, **as useful** to **girder-makers** as to students of Bridge Construction."—("*The Architect*" **on the Second** *Edition.*)

"Of late years the American treatises on Practical and Applied Mechanics have taken the lead . . . since the opening up of a vast continent has given the American engineer a number of new bridge-problems to solve . . . but we look to the PRESENT TREATISE ON BRIDGE-CONSTRUCTION, and the Forth Bridge, to bring us to the front again."—*Engineer.*

"One of the VERY BEST RECENT WORKS on the Strength of Materials and its application to Bridge-Construction. . . . Well repays a careful Study."— *Engineering.*

"An INDISPENSABLE HANDBOOK for the practical Engineer."—*Nature.*

"An admirable account of the theory and process of bridge-design, AT ONCE SCIENTIFIC AND THOROUGHLY PRACTICAL. It is a book such as we have a right **to** expect from one who is himself a substantial contributor to the theory of **the** subject, as well as a bridge-builder of repute."—*Saturday Review.*

"This book is a model of what an engineering treatise ought to be."— *Industries.*

"A SCIENTIFIC TREATISE OF **GREAT** MERIT."—*Westminster Review.*

"Of recent text-books on subjects of mechanical science, there has appeared no one more ABLE, EXHAUSTIVE, or USEFUL than Mr. Claxton Fidler's work on Bridge-Construction."—*Scotsman.*

LONDON: EXETER STREET, STRAND.

ORE & STONE MINING.

BY

C. LE NEVE FOSTER, D.Sc., F.R.S.,

PROFESSOR OF MINING, ROYAL COLLEGE OF SCIENCE; H.M. INSPECTOR OF MINES.

In Large 8vo. With Frontispiece and 716 Illustrations. 34s.

"**Dr. Foster's book was expected** to be EPOCH-MAKING, and it fully justifies such expectation. . . . A MOST ADMIRABLE account of the mode of occurrence of practically ALL KNOWN MINERALS. Probably stands UNRIVALLED for completeness."—*The Mining Journal.*

GENERAL CONTENTS.

INTRODUCTION. Mode of Occurrence of Minerals: Classification; Tabular Deposits, Masses—Examples: Alum, Amber, Antimony, Arsenic, Asbestos, Asphalt, Barytes, Borax, Boric Acid, Carbonic Acid, Clay, Cobalt Ore, Copper Ore, Diamonds, Flint, Freestone, Gold Ore, Graphite, Gypsum, Ice, Iron Ore, Lead Ore, Manganese Ore, Mica, Natural Gas, Nitrate of Soda, Ozokerite, Petroleum, Phosphate of Lime Potassium Salts, Quicksilver Ore, Salt, Silver Ore, Slate, Sulphur, Tin Ore, Zinc Ore. **Faults. Prospecting:** Chance Discoveries — Adventitious Finds — Geology as a Guide to Minerals—Associated Minerals—Surface Indications. **Boring:** Uses of Bore-holes—Methods of Boring Holes: i. By Rotation, ii. By Percussion with Rods, iii. By Percussion with Rope. **Breaking Ground:** Hand Tools—Machinery—Transmission of Power—Excavating Machinery: i. Steam Diggers, ii. Dredges, iii. Rock Drills, iv. Machines for Cutting Grooves, v. Machines for Tunnelling—Modes of using Holes—Driving and Sinking—Fire-setting—Excavating by Water. **Supporting Excavations:** Timbering—Masonry—Metallic Supports—Watertight Linings—Special Processes. **Exploitation:** Open Works:—Hydraulic Mining—Excavation of Minerals under Water—Extraction of Minerals by Wells and Bore-holes—Underground Workings—Beds—Veins—Masses. **Haulage or Transport:** Underground: by Shoots, Pipes, Persons, Sledges, Vehicles, Railways, Machinery, Boats—Conveyance above Ground. **Hoisting or Winding:** Motors, Drums, **and** Pulley Frames—Ropes, Chains, and Attachments—Receptacles—Other Appliances—Safety Appliances—Testing Ropes—Pneumatic Hoisting. **Drainage:** Surface Water —Dams—Drainage Tunnels—Siphons—Winding Machinery—Pumping Engines above ground—Pumping Engines below ground—Co-operative Pumping. **Ventilation:** Atmosphere of Mines—Causes of Pollution of Air—Natural Ventilation—Artificial Ventilation: i. Furnace Ventilation, ii. Mechanical Ventilation—Testing the Quality of Air—Measuring the Quantity and Pressure of the Air—Efficiency of. Ventilating Appliances — Resistance caused by Friction. **Lighting:** Reflected Daylight — Candles—Torches—Lamps—Wells Light—Safety Lamps—Gas—Electric Light. **Descent and Ascent:** Steps and Slides—Ladders—Buckets and Cages—Man Engine. **Dressing:** i. Mechanical Processes: Washing, Hand Picking, Breaking Up, Consolidation, Screening—ii. Processes depending on Physical Properties: Motion in Water, Motion in Air—Desiccation—Liquefaction and Distillation—Magnetic Attraction—iii. Chemical Processes: Solution, Evaporation and Crystallisation, Atmospheric Weathering, Calcination, Cementation, Amalgamation—Application of Processes—Loss in Dressing—Sampling. **Principles of Employment of Mining Labour:** Payment by Time, Measure, or Weight—By Combination of these — By Value of Product. **Legislation affecting Mines and Quarries:** Ownership—Taxation—Working Regulations—Metalliferous Mines Regulation Acts —Coal Mines Regulation Act—Other Statutes. **Condition of the Miner:** Clothing —Housing—Education—Sickness—Thrift—Recreation. **Accidents:** Death Rate of Miners from Accidents—Relative Accident Mortality Underground and Above-ground—Classification of Accidents—Ambulance Training.

"This EPOCH-MAKING work . . . appeals to MEN OF EXPERIENCE no less than to students . . . gives numerous examples from the MINING PRACTICE of EVERY COUNTRY. Many of its chapters are upon subjects not usually dealt with in text books.   Of great interest. . . . Admirably illustrated."—*Berg- und Hüttenmännische Zeitung.*
"This SPLENDID WORK."—*Oesterr. Ztschrft. für Berg- und Hüttenwesen.*

LONDON: EXETER STREET, STRAND.

Cloth, for Office use, 8s. 6d.   Leather, for the Pocket, 8s. 6d.

GRIFFIN'S ELECTRICAL PRICE-BOOK,

FOR THE USE OF

Electrical, Civil, Marine, and Borough Engineers, Local
Authorities, Architects, Railway Contractors, &c., &c.

EDITED BY

H. J. DOWSING,

*Member of the Institution of Electrical Engineers; of the Society of Arts; of the London
Chamber of Commerce, &c.*

GENERAL CONTENTS.

PART I.—PRICES AND DETAILS OF MACHINERY AND APPARATUS.
PART II.—USEFUL INFORMATION CONCERNING THE SUPPLY OF
ELECTRICAL ENERGY; Complete Estimates; Reports, Rules and Regulations, Useful Tables, &c.; and General Information regarding the carrying out
of Electrical Work.

"The ELECTRICAL PRICE-BOOK REMOVES ALL MYSTERY about the cost of Electrical
Power. By its aid the EXPENSE that will be entailed by utilising electricity on a large or
small scale can be discovered.   . Contains that sort of information which is most often
required in an architect's office when the application of Electricity is being considered."—
Architect.

"The value of this Electrical Price-Book CANNOT BE OVER-ESTIMATED. . . . Will
save time and trouble both to the engineer and the business man."—*Machinery.*

GRIFFIN (John Joseph, F.C.S.):

CHEMICAL RECREATIONS: A Popular Manual of Experimental
Chemistry. With 540 Engravings of Apparatus. *Tenth Edition.* Crown
8vo. Cloth. Complete in one volume, cloth, gilt top, 12/6.

Part I.—Elementary Chemistry, 2/.
Part II.—The Chemistry of the Non-Metallic Elements, 10/6.

GURDEN (Richard Lloyd, Authorised Surveyor
for the Governments of New South Wales and Victoria):

TRAVERSE TABLES: computed to Four Places Decimals for every
Minute of Angle up to 100 of Distance. For the use of Surveyors and
Engineers. *Third Edition.* Folio, strongly half-bound, 21/.

** *Published with Concurrence of the Surveyors-General for New South
Wales and Victoria.*

"Those who have experience in exact SURVEY-WORK will best know how to appreciate
the enormous amount of labour represented by this valuable book. The computations
enable the user to ascertain the sines and cosines for a distance of twelve miles to within
half an inch, and this BY REFERENCE TO BUT ONE TABLE, in place of the usual Fifteen
minute computations required. This alone is evidence of the assistance which the Tables
ensure to every user, and as every Surveyor in active practice has felt the want of such
assistance, few knowing of their publication will remain without them."—*Engineer.*

LONDON: EXETER STREET, STRAND.

Griffin's Standard Publications

FOR

ENGINEERS, ELECTRICIANS, ARCHITECTS, BUILDERS,
NAVAL CONSTRUCTORS, AND SURVEYORS.

LONDON: EXETER STREET, STRAND.

In Large 8vo, with Illustrations and Folding-Plates.   10s. 6d.

BLASTING:

A Handbook for the Use of Engineers and others Engaged in Mining, Tunnelling, Quarrying, &c.

By OSCAR GUTTMANN, Assoc. M. Inst. C.E.

Member of the Societies of Civil Engineers and Architects of Vienna and Budapest, Corresponding Member of the Imp. Roy. Geological Institution of Austria, &c.

. Mr. Guttmann's *Blasting* is the ONLY work on the subject which gives at once full information as to the NEW METHODS adopted since the introduction of Dynamite, and, at the same time, the results of MANY YEARS PRACTICAL EXPERIENCE both in Mining Work and in the Manufacture of Explosives. It therefore presents in concise form all that has been *proved* good in the various methods of procedure. The Illustrations form a special and valuable feature of the work.

GENERAL CONTENTS.—Historical Sketch—Blasting Materials—Blasting Powder—Various Powder-mixtures—Gun-cotton—Nitro-glycerine and Dynamite—Other Nitro-compounds—Sprengel's Liquid (acid) Explosives—Other Means of Blasting—Qualities, Dangers, and Handling of Explosives—Choice of Blasting Materials—Apparatus for Measuring Force—Blasting in Fiery Mines—Means of Igniting Charges—Preparation of Blasts—Bore-holes—Machine-drilling—Chamber Mines—Charging of Bore-holes—Determination of the Charge—Blasting in Bore-holes—Firing—Straw and Fuze Firing—Electrical Firing—Substitutes for Electrical Firing—Results of Working—Various Blasting Operations—Quarrying—Blasting Masonry, Iron and Wooden Structures—Blasting in earth, under water, of ice, &c.

"This ADMIRABLE work."—*Colliery Guardian.*
"Should prove a *vade-mecum* to Mining Engineers and all engaged in practical work."
—*Iron and Coal Trades Review.*

With Numerous Illustrations.   Price 12s 6d.

PAINTERS' COLOURS, OILS, AND VARNISHES:
A Practical Manual.

By GEORGE H. HURST, F.C.S.,

Member of the Society of Chemical Industry; Lecturer on the Technology of Painters' Colours, Oils, and Varnishes, the Municipal Technical School, Manchester.

GENERAL CONTENTS.—Introductory—THE COMPOSITION, MANUFACTURE, ASSAY, and ANALYSIS of White Pigments—Red Pigments—Yellow and Orange Pigments—Green Pigments—Blue Pigments—Brown Pigments—Black Pigments—LAKES—Colour and Paint Machinery—Paint Vehicles (Oils, Turpentine, &c., &c.)—Driers—VARNISHES.

"This useful book most successfully combines Theory and Practice . . . will prove MOST VALUABLE. We feel bound to recommend it to ALL engaged in the arts concerned."—*Chemical News.*
"A *practical* manual in every respect. The directions are concise, clearly intelligible, and EXCEEDINGLY INSTRUCTIVE. The section on Varnishes the most reasonable we have met with."—*Chemist and Druggist.*
"A work that is both useful and necessary, from the pen of a writer experienced in more ways than one with the very wide subject with which he deals. VERY VALUABLE information is given."—*Plumber and Decorator.*
"A THOROUGHLY PRACTICAL book, . . . constituting, we believe, the ONLY English work that satisfactorily treats of the manufacture of oils, colours, and pigments."—*Chemical Trades' Journal.*
"Throughout the work are scattered hints which are INVALUABLE to the intelligent reader."—*Invention.*

LONDON: EXETER STREET, STRAND.

COAL-MINING (A Text-Book of):

FOR THE USE OF COLLIERY MANAGERS AND OTHERS ENGAGED IN COAL-MINING.

BY

HERBERT WILLIAM HUGHES, F.G.S.,

Assoc. Royal School of Mines, Certificated Colliery Manager.

SECOND EDITION. *In Demy 8vo, Handsome Cloth. With very Numerous Illustrations, mostly reduced from Working Drawings.* 18s.

"The details of colliery work have been fully described, on the ground that collieries are more often made REMUNERATIVE by PERFECTION IN SMALL MATTERS than by bold strokes of engineering. . . . It frequently happens, in particular localities, that the adoption of a combination of small improvements, any of which viewed separately may be of apparently little value, turns an unprofitable concern into a paying one."—*Extract from Author's Preface.*

GENERAL CONTENTS.

Geology: Rocks—Faults—Order of Succession—Carboniferous System in Britain. Coal: Definition and Formation of Coal—Classification and Commercial Value of Coals. Search for Coal: Boring—various appliances used—Devices employed to meet Difficulties of deep Boring—Special methods of Boring—Mather & Platt's, American, and Diamond systems—Accidents in Boring—Cost of Boring—Use of Boreholes. Breaking Ground: Tools—Transmission of Power: Compressed Air, Electricity—Power Machine Drills—Coal Cutting by Machinery—Cost of Coal Cutting—Explosives—Blasting in Dry and Dusty Mines—Blasting by Electricity—Various methods to supersede Blasting. Sinking: Position, Form, and Size of shaft—Operation of getting down to "Stone-head"—Method of proceeding afterwards—Lining shafts—Keeping out Water by Tubbing—Cost of Tubbing—Sinking by Boring—Kind - Chaudron, and Lipmann methods—Sinking through Quicksands —Cost of Sinking. Preliminary Operations: Driving underground Roads—Supporting Roof; Timbering, Chocks or Cogs, Iron and Steel Supports and Masonry—Arrangement of Inset. Methods of Working: Shaft, Pillar, and Subsidence—Bord and Pillar System— Lancashire Method—Longwall Method—Double Stall Method—Working Steep Seams— Working Thick Seams—Working Seams lying near together—Spontaneous Combustion. Haulage: Rails—Tubs—Haulage by Horses—Self-acting Inclines—Direct-acting Haulage —Main and Tail Rope—Endless Chain - Endless Rope—Comparison. Winding: Pit Frames – Pulleys—Cages—Ropes—Guides—Engines—Drums—Brakes—Counterbalancing— Expansion—Condensation—Compound Engines—Prevention of Overwinding—Catches at pit top—Changing Tubs—Tub Controllers—Signalling. Pumping: Bucket and Plunger Pumps—Supporting Pipes in Shaft—Valves—Suspended lifts for Sinking—Cornish and Bull Engines—Davey Differential Engine—Worthington Pump—Calculations as to size of Pumps—Draining Deep Workings—Dams. Ventilation: Quantity of air required— Gases met with in Mines—Coal-dust—Laws of Friction—Production of Air-currents— Natural Ventilation—Furnace Ventilation—Mechanical Ventilators—Efficiency of Fans— Comparison of Furnaces and Fans—Distribution of the Air-current—Measurement of Air-currents. Lighting: Naked Lights—Safety Lamps—Modern Lamps—Conclusions— Locking and Cleaning Lamps—Electric Light Underground—Delicate Indicators. Works at Surface: Boilers—Mechanical Stoking—Coal Conveyors—Workshops. Preparation of Coal for Market: General Considerations—Tipplers—Screens—Varying the Sizes made by Screens—Belts—Revolving Tables—Loading Shoots—Typical Illustrations of the arrangement of Various Screening Establishments—Coal Washing—Dry Coal Cleaning—Briquettes.

"Quite THE BEST BOOK of its kind . . as PRACTICAL in aim as a book can be . . . touches upon every point connected with the actual working of collieries. The illustrations are EXCELLENT."—*Athenæum.*

"A Text-book on Coal-Mining is a great desideratum, and Mr. HUGHES possesses ADMIRABLE QUALIFICATIONS for supplying it. . . . We cordially recommend the work." —*Colliery Guardian.*

"Mr. HUGHES has had opportunities for study and research which fall to the lot of but few men. If we mistake not, his text-book will soon come to be regarded as the STANDARD WORK of its kind."—*Birmingham Daily Gazette.*

. *Note*—The first large edition of this work was exhausted within a few months of publication.

LONDON: EXETER STREET, STRAND.

WORKS BY

ANDREW JAMIESON, M.Inst.C.E., M.I.E.E., F.R.S.E.,

Professor of Electrical Engineering, The Glasgow and West of Scotland Technical College.

PROFESSOR JAMIESON'S ADVANCED MANUALS.

In Large Crown 8vo. Fully Illustrated.

1. STEAM AND STEAM-ENGINES (A Text-Book on).
For the Use of Students preparing for Competitive Examinations. With **over 200** Illustrations, Folding Plates, and Examination Papers. **TENTH EDITION.** Revised and Enlarged, 8/6.

"**Professor** Jamieson fascinates the reader by his CLEARNESS OF CONCEPTION AND SIMPLICITY OF EXPRESSION. His treatment recalls the lecturing of Faraday."—*Athenæum.*
"The BEST BOOK yet published for the use of Students."—*Engineer.*
"Undoubtedly the MOST VALUABLE AND MOST COMPLETE Hand-book on the subject that now exists."—*Marine Engineer.*

2. MAGNETISM AND ELECTRICITY (An Advanced Text-
Book on). Specially arranged for Advanced and "Honours" Students.

3. APPLIED MECHANICS (An Advanced Text-Book on).
Specially arranged for Advanced and "Honours" Students.

PROFESSOR JAMIESON'S INTRODUCTORY MANUALS.

In Crown 8vo, Cloth. With very numerous Illustrations and Examination Papers.

1. STEAM AND THE STEAM-ENGINE (Elementary Text-
Book on). Specially arranged for First-Year Students. FOURTH EDITION. 3/6.

"Quite the RIGHT SORT OF BOOK."—*Engineer.*
"Should be in the hands of EVERY engineering apprentice."—*Practical Engineer.*

2. MAGNETISM AND ELECTRICITY (Elementary Text-
Book on). Specially arranged for First-Year Students. THIRD EDITION. 3/6.

"A CAPITAL TEXT-BOOK . . . The diagrams are an important feature."—*Schoolmaster.*
"A THOROUGHLY TRUSTWORTHY Text-book. . . . Arrangement as good as well can be. . . . Diagrams are also excellent. . . . The subject throughout treated as an essentially PRACTICAL one, and very clear instructions given."—*Nature.*

3. APPLIED MECHANICS (Elementary Text-Book on).
Specially arranged for First-Year Students. SECOND EDITION. 3/6.

"Nothing is taken for granted. . . . The work has VERY HIGH QUALITIES, which may be condensed into the one word 'CLEAR.'"—*Science and Art.*

A POCKET-BOOK of ELECTRICAL RULES and TABLES.

FOR THE USE OF ELECTRICIANS AND ENGINEERS.
Pocket Size. Leather, 8s. 6d. *Tenth Edition,* revised and enlarged.
(See under *Munro and Jamieson.*)

LONDON: EXETER STREET, STRAND.

"The MOST VALUABLE and USEFUL WORK on Dyeing that has yet appeared in the English language . . . likely to be THE STANDARD WORK OF REFERENCE for years to come."— *Textile Mercury.*

In Two Large 8vo Volumes, 920 pp., with a SUPPLEMENTARY Volume, containing Specimens of Dyed Fabrics. Handsome Cloth, 45s.

A

MANUAL OF DYEING:

FOR THE USE OF PRACTICAL DYERS, MANUFACTURERS, STUDENTS, AND ALL INTERESTED IN THE ART OF DYEING.

BY

E. KNECHT, Ph.D., F.I.C.,

Head of the Chemistry and Dyeing Department of the Technical School, Manchester; Editor of "The Journal of the Society of Dyers and Colourists;"

CHR. RAWSON, F.I.C., F.C.S.,

Late Head of the Chemistry and Dyeing Depatment for the Technical College, Bradford; Member of Council of the Society of Dyers and Colourists;

And RICHARD LOEWENTHAL, Ph.D.

GENERAL CONTENTS.—Chemical Technology of the Textile Fabrics— Water—Washing and Bleaching—Acids, Alkalies, Mordants—Natural Colouring Matters—Artificial Organic Colouring Matters—Mineral Colours —Machinery used in Dyeing—Tinctorial Properties of Colouring Matters— Analysis and Valuation of Materials used in Dyeing, &c., &c.

"This MOST VALUABLE WORK . . . will be widely appreciated."—*Chemical News.*

"This authoritative and exhaustive work . . . the MOST COMPLETE we have yet seen on the subject."—*Textile Manufacturer.*

"The MOST EXHAUSTIVE and COMPLETE work on the subject extant."—*Textile Recorder.*

"The distinguished authors have placed in the hands of those daily engaged in the dye-house or laboratory a work of EXTREME VALUE and UNDOUBTED UTILITY . . . appeals quickly to the technologist, colour chemist, dyer, and more particularly to the rising dyer of the present generation. A book which it is refreshing to meet with."—*American Textile Record.*

LONDON: EXETER STREET, STRAND.

ELECTRO-METALLURGY (A Treatise on):

Embracing the Application of Electrolysis to the Plating, Depositing, Smelting, and Refining of various Metals, and to the Reproduction of Printing Surfaces and Art-Work, &c.

By WALTER G. M'MILLAN, F.I.C., F.C.S.,

Chemist and Metallurgist to the Cossipore Foundry and Shell-Factory; Late Demonstrator of Metallurgy in King's College, London.

With numerous Illustrations. Large Crown 8vo. Cloth 10s. 6d.

GENERAL CONTENTS.—Introductory—Sources of Current—General Condition to be observed in Electro-Plating—Plating Adjuncts and Disposition of Plant—Cleansing and Preparation of Work for the Depositing-Vat, and Subsequent Polishing of Plated Goods—Electro-Deposition of Copper—Electrotyping—Electro-Deposition of Silver—of Gold—of Nickel and Cobalt—of Iron—of Platinum, Zinc, Cadmium, Tin, Lead, Antimony, and Bismuth; Electro-chromy—Electro-Deposition of Alloys—Electro-Metallurgical Extraction and Refining Processes — Recovery of certain Metals from their Solutions or Waste Substances—Determination of the Proportion of Metal in certain Depositing Solutions—Appendix.

"This excellent treatise, . . . one of the BEST and MOST COMPLETE manuals hitherto published on Electro-Metallurgy."—*Electrical Review.*

"This work will be a STANDARD."—*Jeweller.*

"Any metallurgical process which REDUCES the COST of production must of necessity prove of great commercial importance. . . . We recommend this manual to ALL who are interested in the PRACTICAL APPLICATION of electrolytic processes."—*Nature.*

SECOND EDITION. *Enlarged, and very fully Illustrated.* Cloth, 4s. 6d.

STEAM - BOILERS:

THEIR DEFECTS, MANAGEMENT, AND CONSTRUCTION.

By R. D. MUNRO,

Chief Engineer of the Scottish Boiler Insurance and Engine Inspection Company.

This work, written chiefly to meet the wants of Mechanics, Engine-keepers, and Boiler-attendants, also contains information of the first importance to every user of Steam-power. It is a PRACTICAL work written for PRACTICAL men, the language and rules being throughout of the simplest nature.

"A valuable companion for workmen and engineers engaged about Steam Boilers, ought to be carefully studied, and ALWAYS AT HAND."—*Coll. Guardian.*

"The subjects referred to are handled in a trustworthy, clear, and practical manner. . . . The book is VERY USEFUL, especially to steam users, artisans, and young engineers."—*Engineer.*

BY THE SAME AUTHOR.

KITCHEN BOILER EXPLOSIONS: Why

they Occur, and How to Prevent their Occurrence? A Practical Handbook based on Actual Experiment. With Diagrams and Coloured Plate, 3s.

LONDON : EXETER STREET, STRAND.

MUNRO & JAMIESON'S ELECTRICAL POCKET-BOOK.

TENTH EDITION, Revised and Enlarged.

A POCKET-BOOK

OF

ELECTRICAL RULES & TABLES

FOR THE USE OF ELECTRICIANS AND ENGINEERS.

BY

JOHN MUNRO, C.E., & PROF. JAMIESON, M.INST.C.E., F.R.S.E.

With Numerous Diagrams.   Pocket Size.   Leather, 8s. 6d.

GENERAL CONTENTS.

UNITS OF MEASUREMENT.	ELECTRO-METALLURGY.
MEASURES.	BATTERIES.
TESTING.	DYNAMOS AND MOTORS.
CONDUCTORS.	TRANSFORMERS.
DIELECTRICS.	ELECTRIC LIGHTING
SUBMARINE CABLES.	MISCELLANEOUS.
TELEGRAPHY.	LOGARITHMS.
ELECTRO-CHEMISTRY.	APPENDICES.

"WONDERFULLY PERFECT. . . . Worthy of the highest commendation we can give it."—*Electrician.*
"The STERLING VALUE of Messrs. MUNRO and JAMIESON'S POCKET-BOOK."—*Electrical Review.*

MUNRO (J. M. H., D.Sc., Professor of Chemistry,

Downton College of Agriculture):

AGRICULTURAL CHEMISTRY AND ANALYSIS: A PRACTICAL HAND-BOOK for the Use of Agricultural Students.

NYSTROM'S POCKET-BOOK OF MECHANICS

AND ENGINEERING. Revised and Corrected by W. DENNIS MARKS, Ph.B., C.E. (YALE S.S.S.), Whitney Professor of Dynamical Engineering, University of Pennsylvania. Pocket Size. Leather, 15s. TWENTIETH EDITION, Revised and greatly enlarged.

LONDON : EXETER STREET, STRAND.

Demy 8vo, Handsome cloth, 18s.

Physical Geology and Palæontology,

ON THE BASIS OF PHILLIPS.

BY

H A R R Y   G O V I E R   S E E L E Y, F.R.S.,

PROFESSOR OF GEOGRAPHY IN KING'S COLLEGE, LONDON.

With Frontispiece in Chromo=Lithography, and Illustrations.

"It is impossible to praise too highly the research which PROFESSOR SEELEY'S 'PHYSICAL GEOLOGY' evidences. IT IS FAR MORE THAN A TEXT-BOOK—it is a DIRECTORY to the Student in prosecuting his researches."—*Presidential Address to the Geological Society*, 1885, *by Rev. Prof. Bonney, D.Sc., LL.D., F.R.S.*

"PROFESSOR SEELEY maintains in his 'PHYSICAL GEOLOGY' the high reputation he already deservedly bears as a Teacher."—*Dr. Henry Woodward, F.R.S., in the "Geological Magazine."*

"PROFESSOR SEELEY's work includes one of the most satisfactory Treatises on Lithology in the English language. . . . So much **that** is not accessible in other works is presented in this volume, that no **Student of** Geology can afford **to** be without it."—*American Journal of Engineering.*

Demy 8vo, Handsome cloth, 34s.

Stratigraphical Geology & Palæontology,

ON THE BASIS OF PHILLIPS.

BY

R O B E R T   E T H E R I D G E, F.R.S.,

OF THE NATURAL HIST. DEPARTMENT, BRITISH MUSEUM, LATE PALÆONTOLOGIST TO THE GEOLOGICAL SURVEY OF GREAT BRITAIN, PAST PRESIDENT OF THE GEOLOGICAL SOCIETY, ETC.

With Map, Numerous Tables, and Thirty=six Plates.

*** PROSPECTUS *of the above important work—perhaps the* MOST ELABORATE *of its kind ever written, and one calculated to give a new strength to the study of Geology in Britain—may be had on application to the Publishers.*

"No such compendium of geological knowledge has ever been brought together before."—*Westminster Review.*

"If PROF. SEELEY's volume was remarkable for its originality and the breadth of its views, Mr. ETHERIDGE fully justifies the assertion made in his preface that his book differs in construction and detail from any known manual. . . . Must take HIGH RANK AMONG WORKS OF REFERENCE."—*Athenæum.*

LONDON: EXETER STREET, STRAND.

THIRD EDITION. With Folding Plates and Many Illustrations.
Large 8vo. Handsome Cloth. 36s.

ELEMENTS OF METALLURGY:
A PRACTICAL TREATISE ON THE ART OF EXTRACTING METALS FROM THEIR ORES.

By J. ARTHUR PHILLIPS, M.Inst.C.E., F.C.S., F.G.S., &c.,
AND H. BAUERMAN, V.P.G.S.

GENERAL CONTENTS.

Refractory Materials.	Antimony	Iron.
Fire-Clays.	Arsenic.	Cobalt.
Fuels, &c.	Zinc.	Nickel.
Aluminium.	Mercury.	Silver.
Copper.	Bismuth.	Gold.
Tin.	Lead.	Platinum.

*** Many NOTABLE ADDITIONS, dealing with new Processes and **Developments,** will be found in the Third Edition.

"Of the THIRD EDITION, we are still able to say that, as a Text-book of Metallurgy, it is THE BEST with which we are acquainted."—*Engineer.*

"The value of this work is almost *inestimable.* There can be no question that the amount of time and labour bestowed on it is enormous. . . . There is certainly no Metallurgical Treatise in the language calculated to prove of such general utility."—*Mining Journal.*

"In this most useful and handsome volume is condensed a large amount of valuable practical knowledge. A careful study of the first division of the book, on Fuels, will be found to be of great value to every one in training for the practical applications of our scientific knowledge to **any** of our metallurgical operations."—*Athenæum.*

"A work which is equally valuable to the Student as **a Text-book, and to the** practical Smelter **as a** Standard Work of Reference. . . . **The Illustrations** are admirable **examples of** Wood Engraving."—*Chemical News.*

POYNTING (J. H., Sc.D., F.R.S., late Fellow
of **Trinity** College, Cambridge; Professor of Physics, Mason **College,** Birmingham):
THE MEAN DENSITY OF THE EARTH: An Essay to which **the** Adams Prize was adjudged in 1893 in the University of Cambridge. In large 8vo, with Bibliography, Illustrations in the Text, and seven Lithographed Plates. 12s. 6d.

"An account of this subject cannot fail to be of GREAT and GENERAL INTEREST to the scientific mind. Especially is this the case when the account is given by one who has contributed so considerably as has Prof. Poynting to our present state of knowledge with respect to a very difficult subject. . . . Remarkably has Newton's estimate been verified by Prof. Poynting."—*Athenæum.*

POYNTING and THOMSON: TEXT-BOOK
OF PHYSICS. (See under *Thomson*).

LONDON: **EXETER STREET, STRAND.**

WORKS BY

W. J. MACQUORN RANKINE, LL.D., F.R.S.,

Late Regius Professor of Civil Engineering in the University of Glasgow.

THOROUGHLY REVISED BY

W. J. MILLAR, C.E.,

Secretary to the Institute of Engineers and Shipbuilders in Scotland.

I. A MANUAL OF APPLIED MECHANICS:

Comprising the Principles of Statics and Cinematics, and Theory of Structures, Mechanism, and Machines. With Numerous Diagrams. Crown 8vo, cloth, 12s. 6d. THIRTEENTH EDITION.

II. A MANUAL OF CIVIL ENGINEERING:

Comprising Engineering Surveys, Earthwork, Foundations, Masonry, Carpentry, Metal Work, Roads, Railways, Canals, Rivers, Waterworks, Harbours, &c. With Numerous Tables and Illustrations. Crown 8vo, cloth, 16s. NINETEENTH EDITION.

III. A MANUAL OF MACHINERY AND MILLWORK:

Comprising the Geometry, Motions, Work, Strength, Construction, and Objects of Machines, &c. Illustrated with nearly 300 Woodcuts. Crown 8vo, cloth, 12s. 6d. SIXTH EDITION.

IV. A MANUAL OF THE STEAM-ENGINE AND OTHER PRIME MOVERS:

With Numerous Tables and Illustrations, and a Diagram of the Mechanical Properties of Steam. Crown 8vo, cloth, 12s. 6d. THIRTEENTH EDITION.

V. USEFUL RULES AND TABLES:

For Architects, Builders, Engineers, Founders, Mechanics, Shipbuilders, Surveyors, &c. With APPENDIX for the use of ELECTRICAL ENGINEERS. By Professor JAMIESON, F.R.S.E. SEVENTH EDITION. 10s. 6d.

VI. A MECHANICAL TEXT-BOOK:

A Practical and Simple Introduction to the Study of Mechanics. By Professor RANKINE and E. F. BAMBER, C.E. With Numerous Illustrations. Crown 8vo, cloth, 9s. FOURTH EDITION.

⁎⁎⁎ The "MECHANICAL TEXT-BOOK" was designed by Professor RANKINE as an INTRODUCTION to the above Series of Manuals.

LONDON: EXETER STREET, STRAND.

PROF. RANKINE'S WORKS—(*Continued*).

VII. MISCELLANEOUS SCIENTIFIC PAPERS.

Royal 8vo. Cloth, 31s. 6d.

Part I. **Papers** relating to Temperature, Elasticity, **and** Expansion of Vapours, Liquids, **and** Solids. Part II. Papers on Energy and its Transformations. Part III. Papers on Wave-Forms, Propulsion of Vessels, &c. With Memoir by Professor TAIT, M.A. Edited by **W. J.** MILLAR, C.E. With fine Portrait on Steel, Plates, **and** Diagrams.

"No more enduring Memorial of Professor Rankine could be devised than the publication of these papers in an accessible form. The Collection is most valuable on account of the nature of his discoveries, and the beauty and completeness of his analysis. . . . The Volume exceeds in importance any work in the same department published in our time."—*Architect.*

REDGRAVE (Gilbert R., Assoc. Inst. C.E.): CALCAREOUS CEMENTS: Their Nature, Preparation, and Uses, with some observations on Cement Testing. A Practical Hand-Book. (*Griffin's Technological Manuals*).

PETROLEUM:

A Treatise on the Geographical Distribution, Geological Occurrence, Chemistry, Production, **and** Refining of Petroleum; its Testing, Transport, **and Storage,** and **the** Legislative Enactments relating thereto; together with a description of the Shale Oil Industry,

BY

BOVERTON REDWOOD, F.R.S.E., F.I.C., Assoc. Inst. C.E.,

Hon. Corr. Mem. of the Imperial Russian Technical Society; Mem. of the American Chemical Society; Consulting Adviser to the Corporation of London under the Petroleum Acts, &c., &c.

ASSISTED BY

GEO. T. HOLLOWAY, F.I.C., A.R.C.Sc.

In Large 8vo. With Maps and Illustrations.

** SPECIAL FEATURES of Mr. REDWOOD's Work are (1) the hitherto unpublished descriptions of the UNDEVELOPED SOURCES of PETROLEUM in various parts of the world, which the author is in an exceptionally favourable position to give; and (2) Rules for the TESTING, TRANSPORT, and STORAGE of Petroleum—these subjects are fully dealt with from the point of view of LEGISLATION and the PRECAUTIONS which experience in this and other countries has shown to be necessary in the interests of public safety.

Royal 8vo, Handsome Cloth, 25s.

THE STABILITY OF SHIPS.

BY

SIR EDWARD J. REED, K.C.B., F.R.S., M.P.,

KNIGHT OF THE IMPERIAL ORDERS OF ST. STANILAUS OF RUSSIA; FRANCIS JOSEPH OF AUSTRIA; MEDJIDIE OF TURKEY; AND RISING SUN OF JAPAN; VICE-PRESIDENT OF THE INSTITUTION OF NAVAL ARCHITECTS.

With numerous Illustrations and Tables.

THIS work has been written for the purpose of placing in the hands of Naval Constructors, Shipbuilders, Officers of the Royal and Mercantile Marines, and all Students of Naval Science, a complete Treatise upon the Stability of Ships, and is the only work in the English Language dealing exhaustively with the subject.

In order to render the work complete for the purposes of the Shipbuilder, whether at home or abroad, the Methods of Calculation introduced by Mr. F. K. BARNES, Mr. GRAY, M. REECH, M. DAYMARD, and Mr. BENJAMIN, are all given separately, illustrated by Tables and worked-out examples. The book contains more than 200 Diagrams, and is illustrated by a large number of actual cases, derived from ships of all descriptions, but especially from ships of the Mercantile Marine.

The work will thus be found to constitute the most comprehensive and exhaustive Treatise hitherto presented to the Profession on the Science of the STABILITY OF SHIPS.

" Sir EDWARD REED'S ' STABILITY OF SHIPS ' is INVALUABLE. In it the STUDENT, new to the subject, will find the path prepared for him, and all difficulties explained with the utmost care and accuracy; the SHIP-DRAUGHTSMAN will find all the methods of calculation at present in use fully explained and illustrated, and accompanied by the Tables and Forms employed; the SHIPOWNER will find the variations in the Stability of Ships due to differences in forms and dimensions fully discussed, and the devices by which the state of his ships under all conditions may be graphically represented and easily understood; the NAVAL ARCHITECT will find brought together and ready to his hand, a mass of information which he would otherwise have to seek in an almost endless variety of publications, and some of which he would possibly not be able to obtain at all elsewhere."—*Steamship.*

"This IMPORTANT AND VALUABLE WORK . . . cannot be too highly recommended to all connected with shipping interests."—*Iron.*

" This VERY IMPORTANT TREATISE, . . . the MOST INTELLIGIBLE, INSTRUCTIVE, and COMPLETE that has ever appeared."—*Nature.*

"The volume is an ESSENTIAL ONE for the shipbuilding profession."—*Westminster Review.*

RICHMOND (H. Droop, F.C.S., Chemist to the Aylesbury Dairy Company) :

DAIRY CHEMISTRY FOR DAIRY MANAGERS: A Practical Handbook. (*Griffin's Technological Manuals.*)

LONDON: EXETER STREET, STRAND.

Griffin's Metallurgical Series.

STANDARD WORKS OF REFERENCE

FOR

Metallurgists, Mine-Owners, Assayers, Manufacturers,
and all interested in the development of
the Metallurgical Industries.

EDITED BY

W. C. ROBERTS-AUSTEN, C.B., F.R.S.,

CHEMIST AND ASSAYER OF THE ROYAL MINT; PROFESSOR OF METALLURGY IN
THE ROYAL COLLEGE OF SCIENCE.

In Large 8vo, Handsome Cloth. With Illustrations.

Now Ready.

1. **INTRODUCTION to the STUDY of METALLURGY.**
   By the EDITOR. THIRD EDITION. 12s. 6d.

   " **No** English text-book at all approaches this in the COMPLETENESS with which the most modern views on the subject are dealt with. Professor Austen's volume will be INVALUABLE, not only to the student, but also to those whose knowledge of the art is far advanced."—*Chemical News.*
   " INVALUABLE to the student. . . . Rich in matter not to be readily found elsewhere."—*Athenæum.*
   " This volume amply realises the expectations formed as to the result of the labours of so eminent an authority. It is remarkable for its ORIGINALITY of conception and for the large amount of information which it contains. . . . We recommend every one who desires information not only to consult, but to STUDY this work."—*Engineering.*
   " Will at once take FRONT RANK as a text-book."—*Science and Art.*
   " Prof. ROBERTS-AUSTEN'S book marks an epoch in the history of the teaching of metallurgy in this country."—*Industries.*

2. **GOLD (The Metallurgy of).** By THOS. KIRKE ROSE, D.Sc., Assoc. R.S.M., F.I.C., of the Royal Mint. 21s. (See p. 27).

Will be Published at Short Intervals.

3. **COPPER (The Metallurgy of).** By THOS. GIBB, Assoc. R.S.M., F.I.C., F.C.S.

4. **IRON and STEEL (The Metallurgy of).** By THOS. TURNER, Assoc. R.S.M., F.I.C., F.C.S.

5. **METALLURGICAL MACHINERY: the Application of Engineering to Metallurgical Problems.** By HENRY CHARLES JENKINS, Wh.Sc., Assoc. R.S.M., Assoc.M.Inst.C.E., of the Royal Mint.

6. **ALLOYS.** By the EDITOR.

** Other Volumes in Preparation.

LONDON: EXETER STREET, STRAND.

SECOND EDITION, Revised and Enlarged.
In Large 8vo, Handsome cloth, 34s.

HYDRAULIC POWER

AND

HYDRAULIC MACHINERY.

BY

HENRY ROBINSON, M. INST. C.E., F.G.S.,

FELLOW OF KING'S COLLEGE, LONDON; PROF. OF CIVIL ENGINEERING,
KING'S COLLEGE, ETC., ETC.

With numerous Woodcuts, and Sixty-nine Plates.

GENERAL CONTENTS.

Discharge through Orifices—Gauging Water by Weirs—Flow of Water through Pipes—The Accumulator—The Flow of Solids—Hydraulic Presses and Lifts—Cyclone Hydraulic Baling Press—Anderton Hydraulic Lift—Hydraulic Hoists (Lifts)—The Otis Elevator—Mersey Railway Lifts—City and South London Railway Lifts—North Hudson County Railway Elevator—Lifts for Subways—Hydraulic Ram—Pearsall's Hydraulic Engine—Pumping-Engines—Three-Cylinder Engines—Brotherhood Engine—Rigg's Hydraulic Engine—Hydraulic Capstans—Hydraulic Traversers—Movable Jigger Hoist—Hydraulic Waggon Drop—Hydraulic Jack—Duckham's Weighing Machine—Shop Tools—Tweddell's Hydraulic Rivetter—Hydraulic Joggling Press—Tweddell's Punching and Shearing Machine—Flanging Machine—Hydraulic Centre Crane—Wrightson's Balance Crane—Hydraulic Power at the Forth Bridge—Cranes—Hydraulic Coal-Discharging Machines—Hydraulic Drill—Hydraulic Manhole Cutter—Hydraulic Drill at St. Gothard Tunnel—Motors with Variable Power—Hydraulic Machinery on Board Ship—Hydraulic Points and Crossings—Hydraulic Pile Driver—Hydraulic Pile Screwing Apparatus—Hydraulic Excavator—Ball's Pump Dredger—Hydraulic Power applied to Bridges—Dock-gate Machinery—Hydraulic Brake—Hydraulic Power applied to Gunnery—Centrifugal Pumps—Water Wheels—Turbines—Jet Propulsion—The Gerard-Barré Hydraulic Railway—Greathead's Injector Hydrant—Snell's Hydraulic Transport System—Greathead's Shield—Grain Elevator at Frankfort—Packing—Power Co-operation—Hull Hydraulic Power Company—London Hydraulic Power Company—Birmingham Hydraulic Power System—Niagara Falls—Cost of Hydraulic Power—Meters—Schönheyder's Pressure Regulator—Deacon's Waste-Water Meter.

"A Book of great Professional Usefulness."—*Iron.*

** The SECOND EDITION of the above important work has been thoroughly revised and brought up to date. Many new full-page Plates have been added—the number being increased from 43 in the First Edition to 69 in the present. Full Prospectus, giving a description of the Plates, may be had on application to the Publishers.

LONDON: EXETER STREET, STRAND.

GRIFFIN'S METALLURGICAL SERIES.

THE METALLURGY OF GOLD,

BY

T. KIRKE ROSE, D.Sc., A.R.S.M., F.C.S.,

Assistant Assayer of the Royal Mint.

LARGE 8VO, HANDSOME CLOTH, ILLUSTRATED. 21S.

LEADING FEATURES.

1. Adapted for all who are interested in the Gold Mining Industry, **being** free from technicalities as far as possible ; of special value to those engaged **in** the industry—viz., mill-managers, reduction-officers, &c.

2. The whole ground implied by the term "Metallurgy of Gold" has been covered with equal care ; the space is carefully apportioned to **the various** branches of the subject, according to their relative importance.

3. The MACARTHUR-FORREST CYANIDE PROCESS is fully described **for the** first time. By this process over £2,000,000 of gold per annum (at the rate of) is now being extracted, or nearly one-tenth of the total world's production. The process, introduced in 1887, has only had short newspaper accounts given of it previously. The chapters have been submitted to, and revised by, Mr. MacArthur, and so freed from all possible inaccuracies.

4. Among other new processes **not previously described in a text-book are—** (1) The modern **barrel chlorination process, practised with great success in** Dakota, where the Black Hills district is undergoing rapid development owing to its introduction. (2) New processes for separating gold from silver—viz., the **new Gutzkow process, and the Electrolytic process ; the cost** of separation is reduced by them by one-half.

5. A new feature is the description of EXACT METHODS employed in particular extraction works—Stamp-batteries of South Africa, Australia, New Zealand, California, Colorado, and **Dakota;** Chlorination works also, in many parts of **the world ;** Cyanide works of S. Africa and New Zealand. **These accounts are** of special value to practical men.

6. The bibliography is the first made since 1882.

"Mr. Rose gained his experience in the Western States of America, **but he has** secured details of gold-working from ALL PARTS of the world, and these should be of GREAT SERVICE to practical men. . . . The four chapters on *Chlorination*, written from the point of view alike of the practical man and the chemist, TEEM WITH CONSIDERATIONS HITHERTO UNRECOGNISED, and constitute an addition to the literature of Metallurgy, which will prove to be **of** classical value."—*Nature.*

"The most complete description of the chlorination **process** which has yet been published. —*Mining Journal.*

SCHWACKHÖFER and BROWNE:

FUEL AND WATER: A Manual for Users of Steam and Water. By Prof. FRANZ SCHWACKHÖFER of Vienna, and WALTER R. BROWNE, M.A., C.E., late Fellow of Trinity College, Cambridge. **Demy 8vo, with Numerous** Illustrations, 9/.

GENERAL CONTENTS.—Heat and Combustion—Fuel, Varieties of—Firing Arrangements : Furnace, Flues, Chimney — The Boiler, Choice of — Varieties — Feed-water Heaters— Steam Pipes—Water : Composition, Purification—Prevention of Scale, &c., &c.

"The Section on Heat is one **of** the best and most lucid ever written."—*Engineer.*
"Cannot fail to be valuable to thousands using steam power."—*Railway Engineer.*

SEXTON (Humboldt, **F.I.C.**, F.C.S., F.R.S.E., Prof. of Metallurgy, Glasgow and West of Scotland Technical College) :

———METALLURGY (AN ELEMENTARY **MANUAL** OF). With numerous Illustrations. Crown 8vo, extra.

———OUTLINES OF QUANTITATIVE ANALYSIS. For the Use of Students. With Illustrations. FOURTH EDITION. Crown 8vo, Cloth, 3s.

"A COMPACT LABORATORY GUIDE for beginners was wanted, and the want has been WELL SUPPLIED. . . . A good and useful book."—*Lancet.*

———OUTLINES OF QUALITATIVE ANALYSIS. For the Use of Students. With Illustrations. THIRD EDITION. **Crown 8vo, Cloth, 3s. 6d.**

"The work of a thoroughly practical chemist."—*British Medical Journal.*
"Compiled with great care, and will supply a want."—*Journal of Education.*

SHELTON-BEY (W. Vincent, **Foreman** to the Imperial Ottoman Gun Factories, Constantinople) :

THE MECHANIC'S **GUIDE: A Hand-Book for Engineers and** Artizans. With Copious Tables **and Valuable Recipes for Practical Use.** Illustrated. *Second Edition.* **Crown 8vo. Cloth, 7/6.**

SMITH (Robert H., M.Inst.Mech.E., Prof. of Engineering, **Mason Science** College, Birmingham) :

GRAPHIC TABLES for the CONVERSION OF MEASUREMENTS (English and French). 43 Diagrams **for** the Mutual Conversion of Measurements in Different Units of LENGTHS, AREAS, VOLUMES, WEIGHTS, STRESSES, DENSITIES, QUANTITIES OF WORK, HORSE-POWERS, TEMPERATURES, &c. For the Use of Practical Engineers, **Surveyors,** Architects, and Contractors. **4to,** Price 7s. 6d.

Eleventh Edition. Price 18s.

Demy 8vo, Cloth. With Numerous Illustrations, reduced from Working Drawings.

A MANUAL OF
MARINE ENGINEERING:

COMPRISING THE DESIGNING, CONSTRUCTION, AND WORKING OF MARINE MACHINERY.

By A. E. SEATON, M. Inst. C. E., M. Inst. Mech. E., M.Inst.N.A.

GENERAL CONTENTS.

Part I.—Principles of Marine Propulsion.

Part II.—Principles of Steam Engineering.

Part III.—Details of Marine Engines: Design and Cal-

culations for Cylinders, Pistons, Valves, Expansion Valves, &c.

Part IV.—Propellers.

Part V.—Boilers.

Part VI.—Miscellaneous.

"In the three-fold capacity of enabling a Student to learn how to design, construct, and work a modern Marine Steam-Engine, Mr. Seaton's Manual has NO RIVAL as regards comprehensiveness of purpose and lucidity of treatment."—*Times.*

"The important subject of Marine Engineering is here treated with the THOROUGH-NESS that it requires. No department has escaped attention. . . . Gives the results of much close study and practical work."—*Engineering.*

"By far the BEST MANUAL in existence. . . . Gives a complete account of the methods of solving, with the utmost possible economy, the problems before the Marine Engineer."—*Athenæum.*

"The Student, Draughtsman, and Engineer will find this work the MOST VALUABLE HANDBOOK of Reference on the Marine Engine now in existence."—*Marine Engineer.*

SECOND EDITION. With Diagrams. Pocket-Size, Leather. 8s. 6d.

A POCKET-BOOK OF
MARINE ENGINEERING RULES AND TABLES,

FOR THE USE OF

Marine Engineers, Naval Architects, Designers, Draughtsmen, Superintendents and Others.

BY

A. E. SEATON, M.I.C.E., M.I.Mech.E., M.I.N.A.,

AND

H. M. ROUNTHWAITE, M.I.Mech.E., M.I.N.A.

"ADMIRABLY FULFILS its purpose."—*Marine Engineer.*

LONDON: EXETER STREET, STRAND.

By PROFESSORS J. J. THOMSON & POYNTING.

In Large 8vo. Fully Illustrated.

A TEXT-BOOK OF PHYSICS:

COMPRISING

PROPERTIES OF MATTER; HEAT; SOUND AND LIGHT;
MAGNETISM AND ELECTRICITY.

BY

J. H. POYNTING,
SC.D., F.R.S.,
Late Fellow of Trinity College, Cambridge;
Professor of Physics, Mason College,
Birmingham.

AND

J. J. THOMSON,
M.A., F.R.S.,
Fellow of Trinity College, Cambridge; Prof.
of Experimental Physics in the University
of Cambridge.

SECOND EDITION, *Revised and Enlarged.*   *Pocket-Size, Leather, also for Office Use, Cloth,* 12s.

BOILERS, MARINE AND LAND:

THEIR CONSTRUCTION AND STRENGTH.

A HANDBOOK OF RULES, FORMULÆ, TABLES, &C., RELATIVE TO MATERIAL,
SCANTLINGS, AND PRESSURES, SAFETY VALVES, SPRINGS,
FITTINGS AND MOUNTINGS, &C.

For the Use of all Steam-Users.

BY T. W. TRAILL, M. INST. C. E., F. E. R. N.,
Engineer Surveyor-in-Chief to the Board of Trade.

*** In the New Issue the subject-matter has been considerably extended,
and Tables have been added for Pressures up to 200 lbs. per square inch.

"Very unlike any of **the** numerous treatises on Boilers **which** have preceded it. . . . Really useful. . . . Contains **an** ENORMOUS QUANTITY OF INFORMATION arranged in a very convenient form. . . . Those who have to design boilers will find that they can settle the dimensions for any given pressure with almost no calculation with its aid. . . . A MOST USEFUL VOLUME . . . supplying information to be had nowhere else."—*The Engineer.*

"As a handbook of rules, formulæ, tables, &c., relating to materials, scantlings, and pressures, this work will prove MOST USEFUL. The name of the Author is a sufficient guarantee for its accuracy. It will save engineers, inspectors, and draughtsmen a vast amount of calculation."—*Nature.*

"By such an authority cannot but prove a welcome addition to the literature of the subject. . . . We can strongly recommend it as being the MOST COMPLETE, eminently practical work on the subject."—*Marine Engineer.*

"To the engineer and practical boiler-maker it will prove INVALUABLE. The tables in all probability are the most exhaustive yet published. . . . Certainly deserves a place on the shelf in the drawing office of every boiler shop."—*Practical Engineer.*

LONDON: EXETER STREET, STRAND.

WORKS BY DR. ALDER WRIGHT, F.R.S.

FIXED OILS, FATS, BUTTERS, AND WAXES:

THEIR PREPARATION AND PROPERTIES,

And the Manufacture therefrom of Candles, Soaps, and Other Products.

BY

C. R. ALDER WRIGHT, D.Sc., F.R.S.,

Late Lecturer on Chemistry, St. Mary's Hospital School ; Examiner in "Soap" to the
City and Guilds of London Institute.

In Large 8vo. Handsome Cloth. With 144 Illustrations. 28s.

" Dr. WRIGHT's work will be found ABSOLUTELY INDISPENSABLE by every Chemist.
Teems with information valuable alike to the Analyst and the Technical Chemist."—
The Analyst.

"Will rank as the STANDARD ENGLISH AUTHORITY on OILS and FATS for many
years to come."—*Industries and Iron.*

SECOND EDITION. With very Numerous Illustrations. Handsome Cloth, 6s.
Also Presentation Edition, Gilt and Gilt Edges, 7s. 6d.

THE THRESHOLD OF SCIENCE:

Simple and Amusing Experiments (over 400) in Chemistry and Physics.

⁎⁎⁎ To the NEW EDITION has been added an excellent chapter on the
Systematic Order in which Class Experiments should be carried out for
Educational purposes.

"Any one who may still have doubts regarding the value of Elementary
Science as an organ of education will speedily have his doubts dispelled, if he
takes the trouble to understand the methods recommended by Dr. Alder
Wright. The Additions to the New Edition will be of great service to all
who wish to use the volume, not merely as a ' play-book,' but as an instrument
for the TRAINING of the MENTAL FACULTIES."—*Nature.*

"Step by step the learner is here gently guided through the paths of science,
made easy by the perfect knowledge of the teacher, and made flowery by the
most striking and curious experiments. Well adapted to become the TREASURED
FRIEND of many a bright and promising lad."—*Manchester Examiner.*

LONDON: EXETER STREET, STRAND.

WELLS (Sidney H., Wh.Sc., Assoc.Mem.Inst.C.E.,
Assoc. Mem. Inst. Mech. E.; Principal of, and Head of the Engineering
Department in, Battersea Polytechnic Institute, late of Dulwich College):
   ENGINEERING DRAWING AND DESIGN. A Practical
Manual for Engineering Students. With very numerous Illustrations
and Folding-Plate. In Large Crown 8vo.

  VOL. I.—PRACTICAL GEOMETRY, PLANE, AND SOLID. 3s.
  VOL. II.—MACHINE AND ENGINE DRAWING AND DESIGN. 4s. 6d.

"A THOROUGHLY USEFUL WORK, exceedingly well written. For the many Examples and
Questions we have nothing but praise."—*Nature.*
"The Examples are WELL CHOSEN and treated in an able manner. The Illustrations do
GREAT CREDIT to the publishers."—*Science and Art.*
"A CAPITAL TEXT-BOOK, arranged on an EXCELLENT SYSTEM, calculated to give an in-
telligent grasp of the subject, and not the mere faculty of mechanical copying. . . . Mr.
Wells shows how to make COMPLETE WORKING-DRAWINGS, discussing fully each step in the
design."—*Electrical Review.*

Eleventh Annual Issue. Handsome cloth, 7s. 6d.

THE OFFICIAL YEAR-BOOK
OF THE
Scientific and Learned Societies of Great Britain and Ireland.
COMPILED FROM OFFICIAL SOURCES.
*Comprising (together with other Official Information) LISTS of the
PAPERS read during 1893 before all the LEADING SOCIETIES throughout
the Kingdom engaged in the following Departments of Research:—*

§ 1. Science Generally: *i.e.*, Societies occupy-
  ing themselves with several Branches of
  Science, or with Science and Literature
  jointly.
§ 2. Mathematics and Physics.
§ 3. Chemistry and Photography.
§ 4. Geology, Geography, and Mineralogy.
§ 5. Biology, including Microscopy and An-
  thropology.
§ 6. Economic Science and Statistics.
§ 7. Mechanical Science and Architecture.
§ 8. Naval and Military Science.
§ 9. Agriculture and Horticulture.
§ 10. Law.
§ 11. Literature.
§ 12. Psychology.
§ 13. Archæology.
§ 14. Medicine.

"The YEAR-BOOK OF SOCIETIES is a Record which ought to be of the greatest use for
the progress of Science."—*Sir Lyon Playfair, F.R.S., K.C.B., M.P., Past-President
of the British Association.*
"It goes almost without saying that a Handbook of this subject will be in time
one of the most generally useful works for the library or the desk."—*The Times.*
"British Societies are now well represented in the 'Year-Book of the Scientific
Learned Societies of Great Britain and Ireland.'"—(Art. "Societies" in New Edition of
"Encyclopædia Britannica," vol. xxii.)

Copies of the FIRST ISSUE, giving an Account of the History,
Organization, and Conditions of Membership of the various
Societies, and forming the groundwork of the Series, may still be
had, price 7/6. · *Also Copies of the following Issues.*

The YEAR-BOOK OF SOCIETIES forms a complete INDEX TO
THE SCIENTIFIC WORK of the year in the various Departments.
It is used as a ready HANDBOOK in all our great SCIENTIFIC
CENTRES, MUSEUMS, and LIBRARIES throughout the Kingdom,
and has become an INDISPENSABLE BOOK OF REFERENCE to every
one engaged in Scientific Work.